EFFECTIVE LEADERSHIP SKILLS

for Construction Field Leaders

Jason C. McCarty

atp AMERICAN TECHNICAL PUBLISHERS
Orland Park, Illinois 60467-5756

American Technical Publishers, Inc., Editorial Staff

Editor in Chief:
Jonathan F. Gosse
Vice President—Editorial
Peter A. Zurlis
Director of Product Development:
Cathy A. Scruggs
Assistant Production Manager:
Nicole D. Bigos
Technical Editor:
Cathy A. Scruggs
Supervising Copy Editor:
Catherine A. Mini
Copy Editor:
James R. Hein
Cover Design:
Nick G. Doornbos
Art Supervisor:
Sarah E. Kaducak
Illustration/Layout:
Nick G. Doornbos
Sarah E. Kaducak
Digital Media Coordinator:
Adam T. Schuldt
Digital Resources:
Sean C. Grant
Lauren M. Lenoir
Cory S. Butler
Carl R. Hansen
Voice Talent:
Karin L. Gentile
James R. Hein
Robert E. Stickley
Scott C. Bloom
Bradley R. Burian
John C. Shimala
Joe P. Fredianelli
Michael W. Tarasiewicz
Julie M. Welch
Catherine A. Mini
Cory S. Butler
Carl R. Hansen

1 2 3 4 5 6 7 8 9 – 17 – 9 8 7 6 5 4 3 2

Printed in the United States of America

ISBN 978-0-8269-4041-4

This book is printed on recycled paper.

Contents

Contents

3 PLANNING AND ORGANIZING 49

Contents

Contents

5 COMMUNICATION AND MORALE 115

LEARNER RESOURCES

Effective Leadership Skills for Construction Field Leaders includes access to learner resources that enhance and reinforce the content of the textbook.

- Quick Quizzes®
- Flash Cards
- Forms and Documents
- Discussion and Activities
- Leadership Animations
- ATPeResources.com

These online resources can be accessed using either of the following methods.

- Key ATPeResources.com/QuickLinks into a web browser and then enter QuickLink™ code **335421.**
- Use a Quick Response (QR) reader app on a mobile device to scan the QR code located on the opening page of each chapter.

Learner Resources
ATPeResources.com/QuickLinks
Access Code: **335421**

Book Features

Effective Leadership Skills for Construction Field Leaders is packed with proven advice and case studies explaining how skilled tradesworkers transitioning into leadership roles can use new knowledge and strategies to advance their careers. This user-friendly textbook helps skilled construction workers develop the critical leadership skills that result in the efficient and productive completion of projects on time and within budget.

Throughout the textbook, pull quotes emphasize key points about field leadership, and illustrations provide visual reinforcement of leadership concepts. Each chapter opens with an introduction and a list of learning objectives, which are identified by section. In each section of a chapter, "Leading in the Field" scenarios describe a job-site application of leadership. Each section ends with "For Discussion" and "Take Action" items that elicit critical thinking and offer opportunities to apply leadership concepts using role-plays, report completion, and similar activities. Each chapter ends with a "Case Study" that provides very detailed, real-world examples of field leadership in action. Each Case Study includes an opportunity to analyze the leadership concepts presented, read about the actions taken by effective leaders, and share a personal perspective.

Access is provided to a set of online resources that reinforce learning by providing formative quizzes; terminology flash cards; fillable forms and documents; discussion items and activities; plus access to 14 leadership animations that depict appropriate and inappropriate approaches to job-site scenarios.

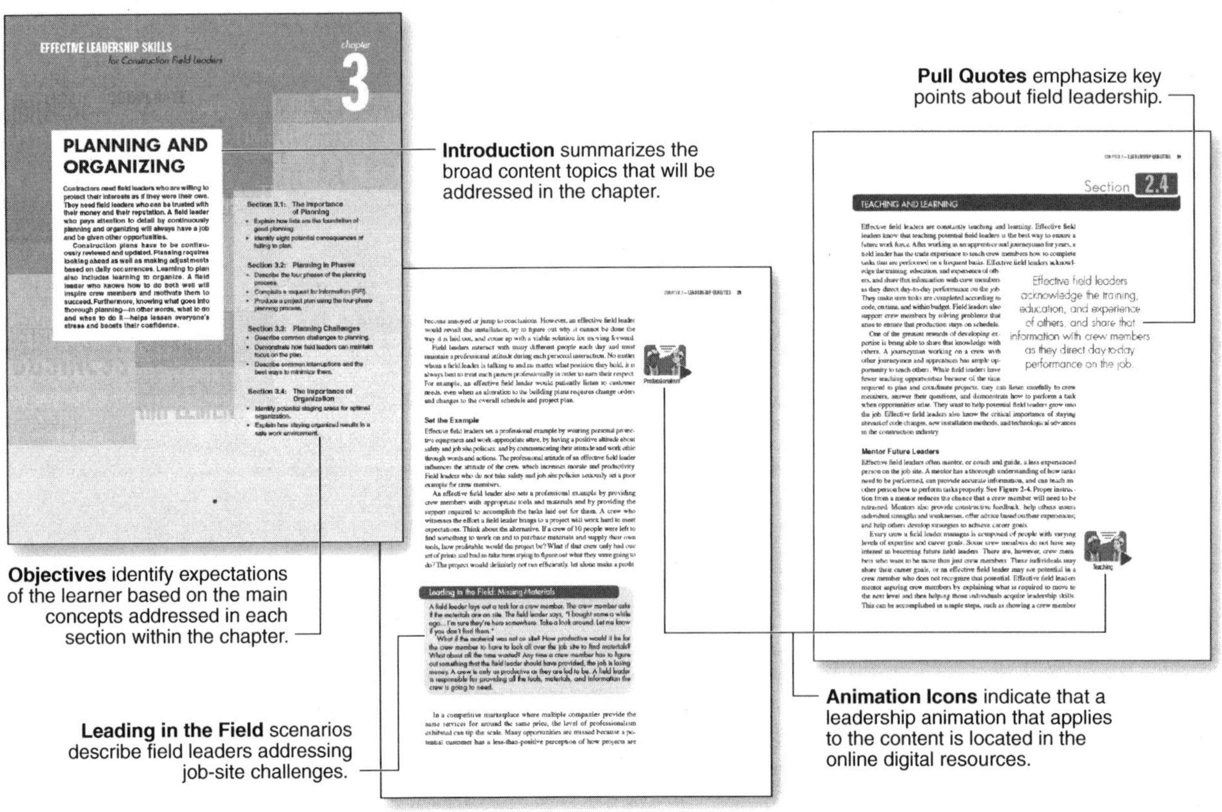

Illustrations provide visual reinforcement of leadership concepts.

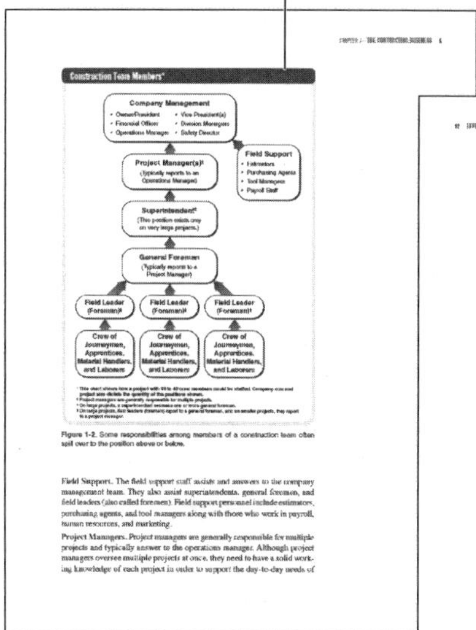

For Discussion items at the end of each section elicit critical thinking and discussion of the content presented.

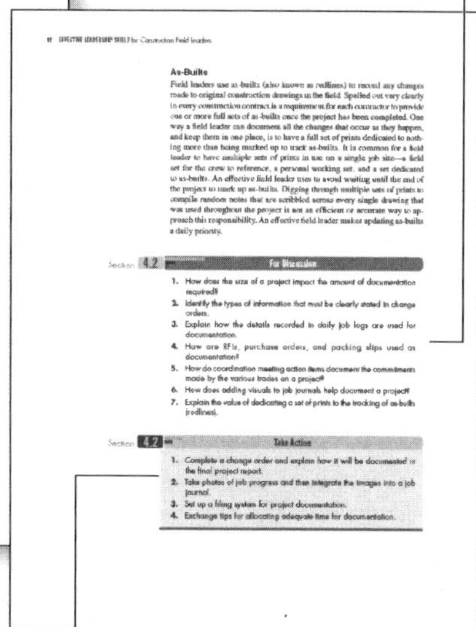

Job-Site Terms provide definitions of terms that are commonly used in construction projects.

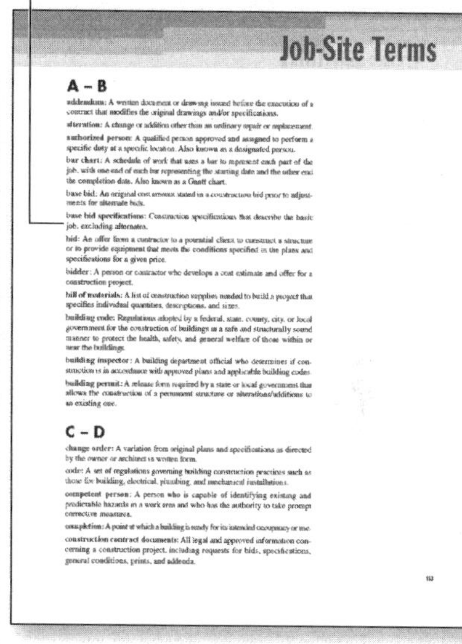

Take Action items at the end of each section provide opportunities to apply leadership concepts using role-plays, report completion, and related activities.

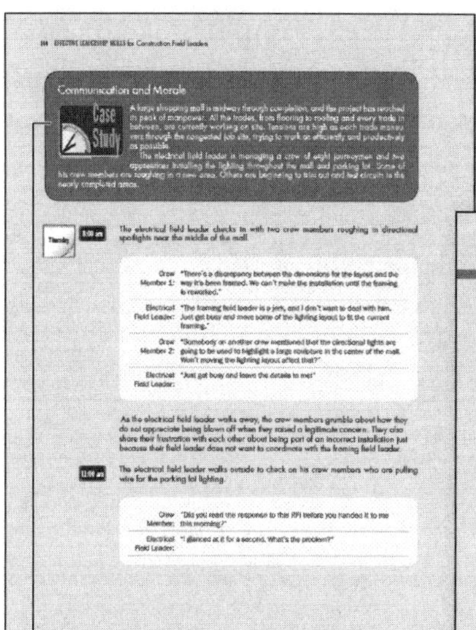

Take A Closer Look provides an opportunity to analyze the leadership concepts presented in the Case Study.

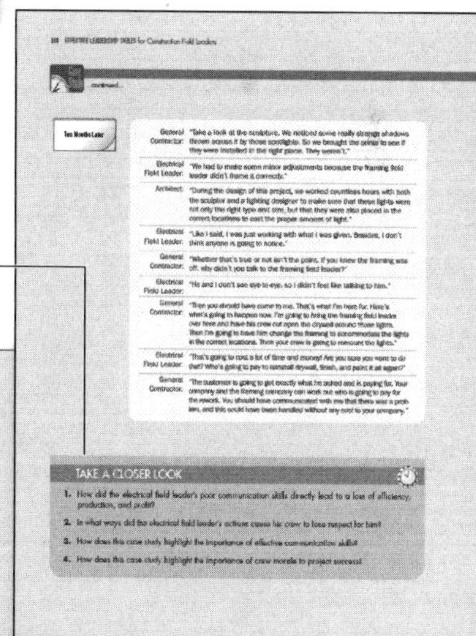

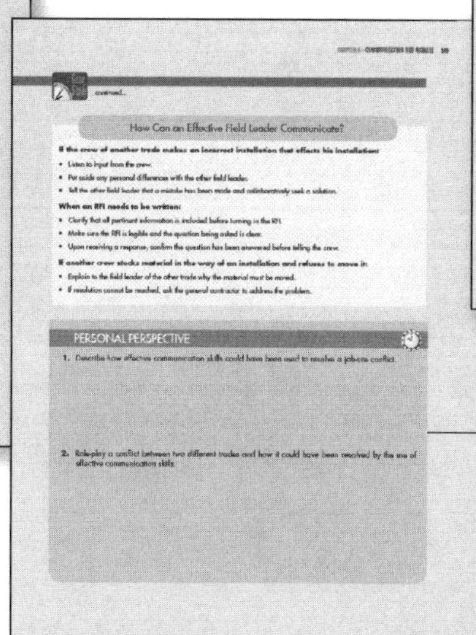

Case Studies provide detailed, real-world examples of field leadership.

Personal Perspective provides an opportunity to reflect on experiences with previous field leaders.

Learner Resources

The *Effective Leadership Skills for Construction Field Leaders* learner resources include the following interactive tools to maximize comprehension and retention.

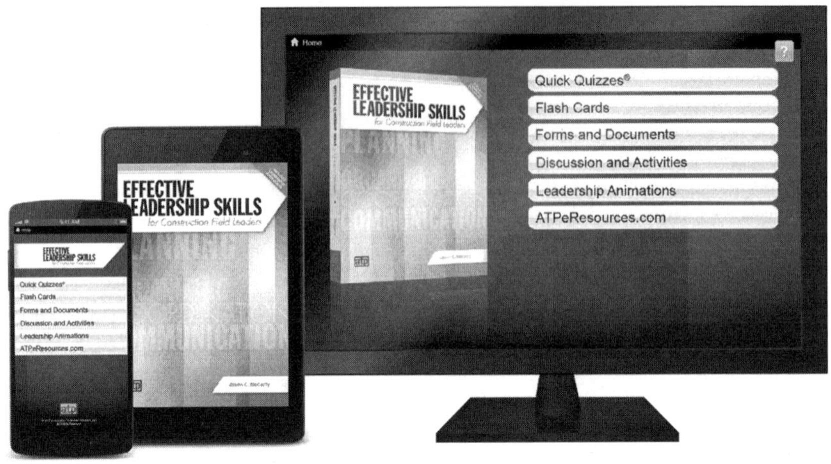

- **Quick Quizzes®** provide 10 interactive multiple-choice questions for each of the 18 sections within the textbook.
- **Flash Cards** provide an interactive review of "Job-Site Terms."
- **Forms and Documents** provide an online method of completing common forms and documents used by field leaders.
- **Discussion and Activities** provide an online method of responding to the "For Discussion" and "Take Action" items listed at the end of each section within every chapter of the textbook.
- **Leadership Animations** present field leadership skills being used on the job site. Each animation contains multiple scenarios to demonstrate both appropriate and inappropriate solutions to job-site interactions that occur between field leaders, crew members, and other tradesworkers.
- **ATPeResources.com** provides access to online reference materials that support continued learning.

To obtain information on related training materials, visit atplearning.com.
—The Publisher

About the Author

Jason C. McCarty is a journeyman inside wireman and a member of the IBEW Local 48 in Portland, Oregon. Jason is a third-generation electrician who started his career as a material handler and laborer in high school. After two years of college, Jason completed a five-year electrical apprenticeship and became a journeyman in both Oregon and Washington by passing each state's electrical licensing exam. He was asked to step up and become a foreman one month after becoming a journeyman in 1998. Since then, Jason has worked and been a leader on a variety of construction projects, including hospitals, schools, manufacturing facilities, industrial plants, corporate high-rises, multiresident high-rises, retail establishments, and other facilities.

In 2003, Jason began working on overseas projects for the State Department as a government contractor. He has worked as an electrician in 22 countries on 5 continents. The diverse nature of this work provided Jason with opportunities to experience different leadership styles that various foremen used to run their projects. These experiences continue to shape his personal approach to leadership and project management.

chapter 1

THE CONTRACTING BUSINESS

Contractors are much too busy to spend large amounts of time managing a project on site. They view field leaders as responsible for keeping projects running smoothly. There is no one within the hierarchy of construction leadership that affects the daily efficiency or production of a crew more than a field leader.

Effective field leaders strive to understand everything they can about contracting—from estimating to the financial considerations that affect a contractor's ability to stay in business. Effective field leaders make informed planning decisions, identify potential problems before they occur, and efficiently lead a crew through the installation process. They combine the skills of a highly qualified tradesperson with a thorough understanding of the construction contracting business.

Section 1.1: The Contractor's View

- Explain how contracting companies bid on construction projects.
- Describe the nine primary groups of people on a construction team.

Section 1.2: The Cost of Doing Business

- Identify the direct and indirect costs of a contracting business.
- Explain the role of risk and cash flow in the contracting business.
- Describe the cost differences between large and small projects.
- Describe five ways that contracting companies can make a profit.

Section 1.3: The Role of Field Leaders

- Describe what makes a person qualified to be a field leader.
- Explain how project planning, crew supervision, and coordination are essential skills of field leadership.

Learner Resources
ATPeResources.com/QuickLinks
Access Code: **335421**

Section 1.1

THE CONTRACTOR'S VIEW

To truly understand how the contracting business works, it is important to understand that it is very rare for a contractor to simply be "given" a project. Contractors typically bid against other contractors to be awarded a project based on their qualifications and the lowest cost proposal. The construction contracting business is very competitive, and it is critical that contractors distinguish themselves from their competitors. No matter what trade or service is provided, there is always another company willing to offer the same product and installation at the same or a lower price. A successful construction contractor provides excellent customer service and delivers quality work at a fair price.

> The construction contracting business is very competitive, and it is critical that contractors distinguish themselves from their competitors.

There are many ways a contractor can be selected to submit a proposal for an upcoming project. One way a bid opportunity can be received is from a current or previous customer. If customers feel they were treated fairly and professionally, then it stands to reason that they would want the same company to complete another project for them in the future.

Leading in the Field: Creative Bidding

ABC Electric, a well-known and respected contractor, has been requested to provide a proposal to wire a new patient tower at a local hospital. The contractor finds out that they are competing against two smaller-but-capable local contractors and another well-established contractor from a neighboring state for the project. The tower project will be the largest of its kind in the area over the next two years, so the competition will be stiff.

In order to be awarded the project, ABC Electric knows they have to be creative in their approach. The first thing ABC does after receiving the construction documents is carefully look them over and search for ways in which they can increase efficiency and production. During this process, their most trusted project managers, estimators, and field leaders work together to develop a realistic plan. The management team formulates a method of installation that will save an average of 19 labor hours wiring each patient room. The new patient tower has a total of 218 rooms, allowing ABC Electric to remove 4142 labor hours from their cost proposal. The team also negotiates an additional 2% savings off the discount they already receive from their largest supplier, on the condition that the vendor will be the sole provider of electrical materials for the entire project. By using a creative and collaborative approach to lower their bid price, ABC Electric is awarded the tower project and earns a reasonable profit.

Competing for Projects

All contracting businesses are in direct competition with companies that operate in the same geographical area. **See Figure 1-1.** However, there are ways that contractors can set themselves apart. Many successful and customer-preferred contractors do this by collaborating with their most experienced estimators, project managers, and field leaders to establish new methods of addressing common problems. Their focus is to increase efficiency and production without sacrificing the quality of craftsmanship the customer is paying for. Contractors who can master this concept of value engineering without giving up quality are more likely to become the preferred provider of services in the market area they serve.

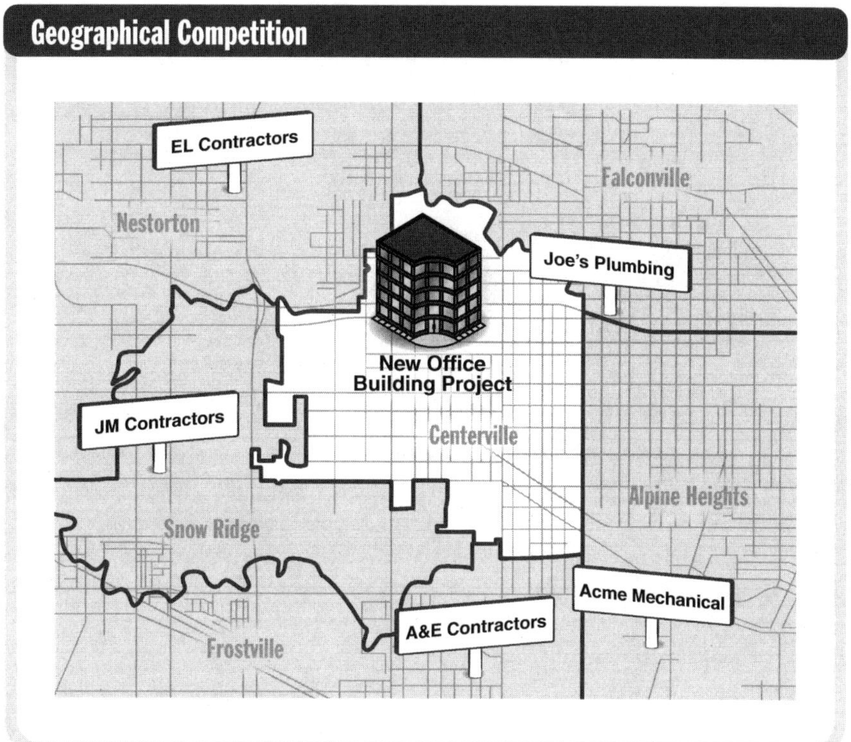

Figure 1-1. All contracting businesses are in direct competition with companies that operate in the same geographical area.

Once a contractor receives a request for proposal (RFP), a complete set of prints and project specifications are reviewed. This is where having experienced estimators and project managers are invaluable to the contractor. A team that estimates and envisions a path from start to finish must draw on their technical expertise and career experience to make educated decisions about how to approach a project. During the estimating process, it is crucial that every aspect of the installation is taken into consideration. Estimating is both a science and an art. No matter how thoroughly the estimating team deciphers the construction documents while preparing a project proposal, there is no way to know for sure if a project has been bid correctly until midway through or even later in the building process.

In most cases, especially on large projects, an RFP will be offered to several contractors, and they all compete for the opportunity to build the project. Once the proposals are submitted, the customer reviews each offer and carefully looks for obvious omissions. Once that step has been completed, the customer selects a contractor, which is most often the one who offered the lowest bid. The moment a contractor is awarded a project (especially a large one), the management team turns their thinking toward finding the most logical, efficient, and productive path to completion. And near the top of the decision list is which field leader(s) to hire to manage the job site.

Some contractors bid too aggressively in an effort to undercut the competition. Reducing the total cost of materials and labor hours by cutting corners on an installation does not provide longevity in the contracting business. Once a company earns a reputation for making substandard installations, that company will be quickly removed from the list of contractors invited to bid on new projects. However, companies that maintain a high level craftsmanship and search for innovative solutions will continue to receive bid opportunities.

The Construction Team

When a contractor is awarded a project, everyone involved from the managers to the field leaders is ecstatic because their hard work on the proposal paid off. They spent weeks, even months, doing take-offs and calculations and sending out equipment and material lists for quotes. Then they sharpened their pencils to make sure they put the right number down. However, shortly after the euphoria of winning wears off, everyone on the team who worked so hard to put the proposal together begins to wonder, "How low were we? Did we leave something out? Will the field leader(s) be able to execute our plan?"

The next step, the actual building of the project, is even more important than the proposal that won it. Now the contractor must turn the proposal into reality. A determined effort from everyone on the construction team, from top to bottom, will be required. It is difficult for a contractor to strictly define roles for construction team members. Some responsibilities among construction team members may overlap. **See Figure 1-2.** Communication is critical to make sure nothing is left out. If a project is not carefully managed, it is very easy for things to slip through the cracks. Following is some general information about the basic responsibilities of a construction team.

Company Management. Leading a contracting business is company management. This group of people establishes the company identity and sets the example for those in their employ. Company management often includes the owner (or president), vice president(s), financial officer, division managers, operations manager, and safety director.

Construction Team Members*

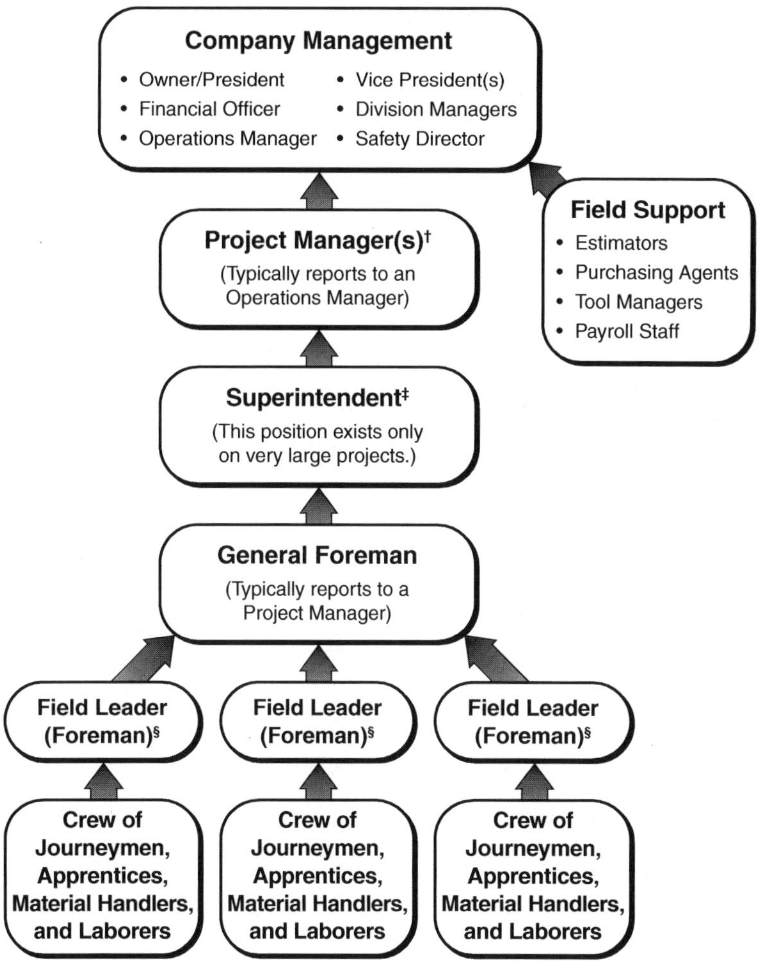

*** This chart shows how a project with 30 to 40 crew members could be staffed. Company size and project size dictate the quantity of the positions shown.**

† Project managers are generally responsible for multiple projects.

‡ On large projects, a superintendent oversees one or more general foremen.

§ On large projects, field leaders (foremen) report to a general foreman, and on smaller projects, they report to a project manager.

Figure 1-2. Some responsibilities among members of a construction team may overlap.

Field Support. The field support staff assists and answers to the company management team. They also assist superintendents, general foremen, and field leaders (also called foremen). Field support personnel include estimators, purchasing agents, and tool managers along with those who work in payroll, human resources, and marketing.

Project Managers. Project managers are generally responsible for multiple projects and typically answer to the operations manager. Although project managers oversee multiple projects at once, they need to have a solid working knowledge of each project in order to support the day-to-day needs of those who work for them. Project managers accomplish this difficult task by maintaining up-to-date prints and project documents for every project. Any reference materials a project manager uses in support of the team needs to accurately match those of each field leader and each customer, and they must be easily accessible for quick reference and verification.

Superintendents. Depending on company size or project size, the definition of the superintendent position varies from contractor to contractor. In some cases, the title of superintendent is given to a person who presides at the highest level of authority on larger projects. This person oversees a staff of general foremen and field leaders that large jobs require. In other cases, a contractor might define a field superintendent as the person who works within the office but supervises every employee in the field.

General Foremen. A general foreman is typically responsible for multiple field leaders (foremen) and holds field leaders accountable for their individual crews. A general foreman remains in constant communication with all field leaders and makes it clear how they will collectively meet any challenges from the beginning to the end of the project.

A general foreman remains in constant communication with all field leaders and makes it clear how they will collectively meet any challenges from the beginning to the end of the project.

Because general foremen attend many more coordination meetings than field leaders, they must be diligent in thoroughly communicating details down the chain of command. It is crucial that a general foreman meet with field leaders regularly to verify that everyone is aware of the upcoming phases of the project. Frequent meetings ensure that future phases of the project are not being ignored and provide an opportunity for the general foreman to address issues and find solutions before they hinder job progress.

Field Leaders (Foremen). A field leader's priorities revolve around the planning, organization, and overall direction for the project and crew on the job site. A crew's ability to work at maximum efficiency and production is only possible when they know exactly what needs to be done and can find what they need to do their jobs. A field leader can only be effective when the crew understands the vision for the project, which is accomplished by consistently communicating expectations.

During nearly all phases of a project, the field leader sees and interacts with the customer the most. Customers do not just buy into what a company has to offer. They also buy into what the field leaders bring to the projects. On large projects, each field leader will more than likely answer to a general foreman. On smaller projects, field leaders will generally report to a project manager who would, on very rare occasions, be on site full time.

Crew Members. Crew members include every journeyman, apprentice, material handler, and laborer working on a construction project. The field leader interacts with each crew member on a daily basis regarding every task, from job site organization and material distribution throughout the site to the building of installations.

Architects and Engineers. Architects and engineers help field leaders navigate through the building process successfully. These construction professionals may not be on the actual job site but still communicate frequently with field leaders via phone, email, and written responses to requests for information, and at some coordination meetings.

Vendors. On larger projects, the project management team will typically sit down with all the vendors who will supply materials for the project. Contact persons are identified and parameters set for what is expected. The contractor provides upfront quantities for any long-lead or special-order items in addition to the dates they will be needed on the job site. The vendors are also informed of any special instructions, including delivery windows, driver requirements for the job site, and any other unique challenges a project may present. These meetings are important, because they create a record of what has been agreed to before the pressures of the project set in. Vendors who understand the challenges faced on the job site consistently provide a high level of customer service and often become long-term partners with contractors and individual field leaders alike.

Section 1.1 For Discussion

1. Why is the contracting business so competitive?
2. How do contracting companies bid on construction projects?
3. Describe the nine groups of people on a construction team.

Section 1.1 Take Action

1. Role-play two competing companies discussing why they are more qualified to win a new office tower project.
2. Role-play a contractor and a field leader (foreman) discussing why they should work together on an upcoming project that the company is bidding on.

Section 1.2

THE COST OF DOING BUSINESS

Every construction contracting business carries financial burdens that remain hidden from those who work on the job site. These costs are expenses that allow a company to keep its doors open. Each contractor faces similar dilemmas of finding field leaders who are experienced enough to make sound financial decisions regarding the projects they are asked to run. Understanding the financial aspects of the company is vital to making the best decisions possible on the job site.

Understanding the financial aspects of the company is vital to making the best decisions possible on the job site.

Most financial reports include material costs incurred, labor hours used, and the total labor hours still available based on what was bid. Companies often ask for the field leader's input to help determine the financial status of a project and also forecast the financial needs for months to come. Effective field leaders take financial considerations into account each time they make a decision on a project. However, profitability can be a confusing concept when only viewed on a project-by-project basis.

For example, a field leader might say that a project was "profitable" based on a contract amount of $100,000 because only $97,000 was spent on tools, materials, labor, and job-related expenditures. In the field leader's mind, the company made a $3,000 profit. The problem is that the field leader did not factor in all the company's expenses before deciding whether a profit was made. If every project that a contractor was awarded finished with a 3% gross profit margin like the example mentioned, that contractor would not be in business for very long.

Contractors must spend a certain amount of money to keep the lights on in the office. These overhead costs can add up, and every contractor must overcome them in order to have a financially viable company. Following are the types of hidden costs that a contractor incurs in addition to the cost of tools, materials, and labor.

Direct and Indirect Costs

Every project incurs both direct and indirect costs. Direct costs are the costs incurred to actually build a project. Direct costs include but are not limited to the following:

- Permit fees
- Field offices and equipment
- Field vehicles
- Field labor, including fringe benefits
- Material costs
- Equipment rental and insurance
- Storage units and space

All contracting companies also have indirect costs, such as operations overhead and labor overhead, that are relative to the size of the business. **See Figure 1-3.** In some cases, a portion of these indirect costs may be charged to the project. One example of indirect costs is the salaries, wages, and benefits of the employees and departments that cannot be charged directly to a specific project. A field leader may forget about the support staff in the company office because they are not on the job site. It is not unusual for a field leader to only visit the company office once or twice a year, if at all. Each time a field leader sends in time cards, writes purchase orders, or calls the office to request more crew members be sent to the job site, a task is added to the support staff's workload. Support staff is critical to the success of every project. Field leaders who run efficient projects help the company offset these indirect costs.

Indirect Costs

Operations Overhead

Operations overhead includes but is not limited to the following:

- Lease/rent/purchase
- Loan interest
- Utilities
- Vehicles, gas, insurance, and maintenance
- Phone/cell service
- Business insurance
- Business licenses
- Bonding
- Legal fees
- Office equipment and maintenance
- Office supplies

Labor Overhead

Labor overhead includes the salaries, wages, and benefits of employees and departments that cannot be charged directly to a specific project. These employees and departments include but are not limited to the following:

- Owner(s)
- Superintendents*
- Managers*
- General Foremen*
- Estimators*
- Purchasing
- Marketing
- Dispatch
- Payroll/Collections
- Human Resources

*In some cases, a portion of the indirect labor costs may be charged to the project.

Figure 1-3. All contracting companies have indirect costs, such as operations overhead and labor overhead, that are relative to the size of the business.

Financial Risk

When a contractor takes on a project, there is no guarantee that it will be profitable. The nature of construction contracting makes earning a profit on every single project a huge risk. Any profits a company stands to make are not only delayed during the completion of every project, but are purely hypothetical until every requirement of the contract has been completed to the customer's satisfaction and the company receives the final payment.

Many factors play into whether or not a project makes a profit. For example, it only takes overlooking a few crucial items during the estimating process for a contractor to unknowingly win a project that is destined to lose money before it starts. Sometimes a contractor hires a person in a field leadership position who does not fully understand what it takes to do the job. Even if a field leader has the best intentions, a lack of experience in planning and coordinating a project or managing a crew can result in a financial loss for the company. Although no one sets out to take on projects that have little-to-no chance of making money, it is bound to happen the longer a contractor stays in business.

Long-term success in the construction industry is only possible when contractors trust field leaders to take financial risk into account when making project-related decisions that protect the company's interests. Effective field leaders make the company's primary objective their primary objective and do the best job they can on every project.

Long-term success in the construction industry is only possible when contractors trust field leaders to take financial risk into account when making project-related decisions that protect the company's interests.

Cash Flow

Generally speaking, when contractors are awarded a project, there are no up-front payments. Companies typically do not receive payment for 30 to 90 days after a project-related expense is incurred. This means that due to the initial costs involved with estimating a project and getting that project underway, contractors either have to operate using the working capital they have earned from previous projects or borrow money from a lender. The contractor signs an agreement with the customer or the general contractor that spells out exactly how much can be billed and when throughout the length of the project. In almost all cases, contractors purchase materials and pay their employees in advance of when they will be reimbursed for expenses.

During the first month, the contractor does not bill the customer at all. During the second month, the contractor bills the customer based on the expenses incurred during the first month, including material, labor, administrative expenses, and a small percentage for profit. During the third month, the contractor expects to receive payment for what was billed for the first month of work. And by the time the payment for what was done during the first month is received, the project has probably ramped up, and more tools and materials are needed. Therefore, the payment for the first month will not cover the new expenses of the third, and the cycle repeats. During an entire project, the contracting company is always behind relative to the money it puts out versus the money it has been paid. **See Figure 1-4.** As a project winds down, although the payments received become larger than the expenses incurred, the contractor may still be in the red.

At the end of a project, the customer can hold the final payment back if they are not fully satisfied that the contractor has fulfilled all commitments. Generally, this final payment is where any profits are finally realized. This makes the end of a project very important. Any loose ends or punch-list items that have not been finished are the very items that can keep the company from receiving its final payment.

Cash Flow

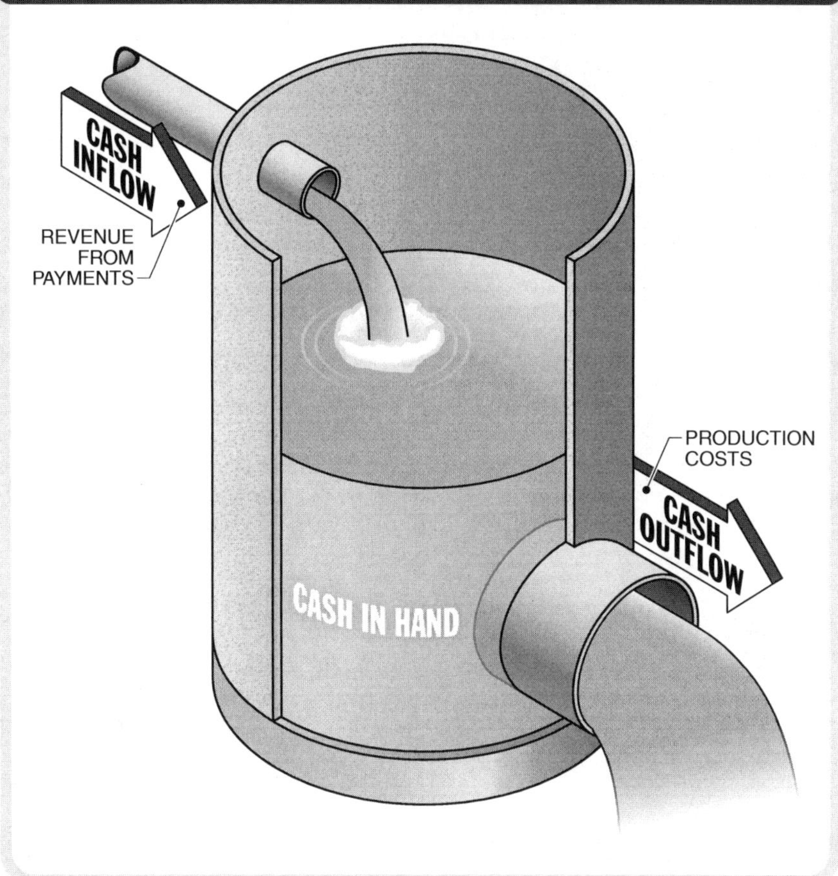

Figure 1-4. During an entire project, the contracting company is always behind relative to the money it puts out versus the money it has been paid.

Small and Large Projects

Contractors accumulate profits through the volume of work their company performs. This means that every single field leader must strive to manage an efficient and profitable project, regardless of the size. For every project that does well, there is always going to be a project that does not. The concept of making a profit based on volume does not work if field leaders do not run efficient and productive jobs.

For example, there is very little room for error, rework, or other oversights that result from poor planning and organization on small projects. That is why field leaders need to think of everything ahead of time to avoid constant starts and stops. When a small project is not run efficiently and requires more labor hours than were bid for the project, the percentage of loss increases dramatically for each wasted labor hour. Small projects do not offer a field leader the ability to ramp up slowly the way large projects do.

One advantage of large projects is the option to negotiate the prices paid for large quantities of items. Since price breaks are sometimes available on volume purchases, the company has the option of getting quotes from multiple vendors. However, larger projects tend to present more leadership challenges than smaller projects. For example, a large project may require a general foreman to be responsible for three very different-minded field leaders. One field leader may be very organized and always have what the crew needs. The second field leader may not always focus on the tools and materials portion of planning. And the third field leader may plan only on the fly. These three field leaders and their crews will likely argue over tools and materials. The field leader who planned may constantly be lending tools and materials to the other two. This same field leader's crew will likely get frustrated because they are always left short—even though they were better prepared. When tools and materials are in short supply, frustration diminishes crew morale and lowers efficiency. Effective field leadership is key to having large projects meet schedules and stay within budget.

Leading in the Field: Viewing Losses

When a project that is bid for 100 hours goes over by 10 hours, it is 10% in the red. But when a project bid for 1000 hours goes over by 10 hours, it is only 1% in the red. Though the loss is the same, the impact in terms of how the company sees the loss is different. A field leader who runs several jobs in a row at 10% over for labor will not likely be in high demand. On the other hand, a field leader who runs jobs only 1% over for labor will likely be given additional opportunities.

How Profit Is Made

Contractors provide field leaders with an avenue to use their knowledge and skills to provide for their families. Every time a contractor goes out of business, field leaders lose a place to find employment. Understanding how contractors make a profit is an important step in becoming an effective field leader. Material and labor mark-ups, time-and-materials projects, change orders, prefabrication, and production all have a direct impact on how profit is made.

Material and Labor Mark-Ups. Generally speaking, contractors mark up the materials they purchase for a project. This means they add a cost that is passed on to the customer, not just for buying the materials but for handling them as well. This is meant to cover things such as the gas for the vehicle used to transport the materials. There are also additional indirect labor costs related to the purchasing and handling of materials as well as the cost involved in processing the billing of the materials, which is completed by support staff.

Although a company may show a mark-up of 10% on its material purchases, this does not mean it will clear a 10% profit on that mark-up. Sometimes on large projects, customers may choose to purchase the big-ticket items themselves, essentially removing the contractor's ability to recover any of the costs related to handling the materials or equipment.

When bidding against other contractors, labor mark-up becomes a primary focal point. Contractors often have many different rates that they charge for the exact same employees depending on the project. Say for instance that a contractor is bidding a very large job with thousands of labor hours. The contractor may lower the hourly rate applied to the bid to be more competitive during the bidding phase and then hope to make up some of that by picking up efficiencies over the length of the project. On the other hand, if a contractor is called to bid on a small job that only requires one employee to be on site for four hours, the contractor would charge a much higher rate. Just like materials, the mark-up of labor is included in part to earn a profit. But even more so, it is used to cover the indirect costs a company incurs for just opening its doors every day.

Time-and-Materials Projects. A time-and-materials project is one where the customer agrees to pay the contractor based on how many hours it actually takes to complete the project and how much material is used. This may sound like the contractor is being handed a blank check, but that is not the case. Most of the time the customer writes a "not to exceed" clause into the contract, which means that the contractor has most likely provided a budgetary number up front that the customer has taken into consideration before agreeing to proceed. As with material mark-ups, there are administrative costs related to this kind of project, including a considerable amount of paperwork. In order to protect their interests, customers often require documentation regarding labor hours and material quantities used in order to authenticate that they are only paying for what is actually done.

Change Orders. Contractors often pick up extra work from the customer in the form of change orders throughout the course of a project. This work includes things that a customer adds or items that were omitted on the original drawings on which the contractor based the original bid price. These changes are priced per change and are usually more profitable than the actual contract work. The mark-up on a change order is generally higher than the standard mark-up that is often lowered during the estimating process in order to win a project. Another reason that change orders are more profitable is the fact that a crew will likely still be on site when a change order is issued, so there will be no additional costs involved with set up, mobilization, or demobilization.

Prefabrication. During the estimating process or at the start of a project, contractors seek places where they may be able to reduce labor hours through efficiencies such as prefabrication. For example, there are times when it is more efficient and productive to build large quantities of similar items in a shop, where they can be produced in a controlled and consistent environment. After production, these items are transported to the job site to be installed. Prefabrication helps the contractor bring down the per-unit cost by not having workers build items one by one.

Prefabrication helps the contractor bring down the per-unit cost by not having workers build items one by one.

There are a couple of advantages gained by prefabrication. First is the ability to order and stock larger amounts of material, potentially increasing the possibility of obtaining a discount from the supplier. The other is the ability to reduce labor hours by building items in an assembly-line fashion. Without the need to frequently relocate workstations on site to allow for other trades to work in one area or another, many setbacks in production can be avoided.

When deciding whether or not prefabrication would be beneficial for a project, the field leader must determine if prefabricating the item will require a lot of adjustment or need to be taken apart on site, which may end up costing more labor hours than would be gained. In addition, there are also costs associated with transporting and handling prefabricated systems as they are moved onto the job site. It is important to consider all the factors related to a prefab scenario before moving forward, as sometimes what looks like a good idea on paper may not be.

Production. At the heart of what really makes or breaks a contractor's chance of making a profit is production. Production is the number one variable that can increase or decrease the profitability of a project.

Hypothetically speaking, if a contractor bids a project for 1000 hours of labor, and it takes exactly 1000 hours to complete the project, then the only profit made is the mark-up that is left after all overhead costs are removed. If that same project is finished using only 900 labor hours, then 100 labor hours plus the mark-up of those 100 hours are left on the table to help pay the company's overhead, and that could earn a very nice profit. In theory, if a project were to use exactly the amount of labor hours for which it was bid and the material costs were exactly as estimated, the contractor would likely make a very small profit. But this type of risk versus reward is not why contractors go into business.

Field leaders may not be able to prevent an expensive piece of equipment from being left out of an estimate, but they can always affect the efficiency and productivity of the crew. Effective field leaders who plan well and manage an efficient crew can minimize the loss from a project that loses money. Success as an effective field leader is determined by the ability to run projects with the highest possible level of efficiency every single time. In the contracting business, it is the combined profits and losses of the projects undertaken that define success.

Section 1.2 For Discussion

1. Explain how direct costs differ from indirect costs.
2. What risk is involved in the contracting business?
3. Describe the typical cash flow process for a construction project.
4. What are the cost differences between small and large projects?
5. Describe five ways that contracting companies can make a profit.

Section 1.2 Take Action

1. Make a list of indirect costs associated with a construction project.
2. Role-play how using prefabricated items resulted in a contractor making a profit.

Section 1.3

THE ROLE OF FIELD LEADERS

The role of a field leader can best be defined as an individual who is the principal planner, organizer, safety leader, decision maker, problem solver, documentarian, communicator, and teacher on a construction project. To rise to the highest levels of construction management, effective field leaders must become proficient in all of these roles. The field leader is the person who will make the most important decisions once a project is underway. The field leader is the only company representative present on site for every single issue that arises. If a field leader does not fully understand the company's business goals, the decisions made on the job site may negatively affect the company's success. After all, the company, not the field leader, shoulders the financial burden of any project that goes over budget.

The role of a field leader can best be defined as an individual who is the principal planner, organizer, safety leader, decision maker, problem solver, documentarian, communicator, and teacher on a construction project.

A construction management team's primary involvement with a project and a customer happens during the estimating and bidding process. Once that step is complete, it is the field leader's responsibility to fulfill every commitment that was made. The job of managing people in the construction industry is challenging, because each project varies in geographical location and technical specifications. This means that every project will require field leaders to make intelligent decisions to overcome unique production challenges.

Qualified to Lead

Ironically, many owners started their contracting businesses based on their own knowledge, abilities, approach to running projects, and enjoyment in taking care of a customer's needs. However, once they became owners, their roles as hands-on leaders during actual construction disappeared. Choosing the right people to carry out the vision is a vital part of being a successful contractor.

The decision-making process each contractor uses when choosing field leaders varies, but there are certain qualities they look for. First, a field leader must demonstrate an above-average knowledge of all facets of the trade. Second, the management team must believe that the new field leader is capable of setting a consistent example that a crew would be willing to follow. A third quality they often consider is whether or not the prospective field leader has the ability to motivate people to work hard. Lastly, the company must believe the person is trustworthy enough to keep them informed of all the financial decisions made that affect the outcome of a project. Without being able to be at the job site every single day to see what is happening, owners and managers of contracting businesses are in the precarious position of trusting field leaders to keep them informed of daily activities.

Leading in the Field: The New Field Leader

A contractor promotes an outstanding crew member to the role of field leader. The crew member has always been reliable and has a good work ethic, and the contractor assumes those qualities will translate into excellent leadership skills. The new field leader has worked for qualified leaders in the past but has never been given the opportunity to witness firsthand what a day in the life of an effective field leader entails.

In the early stages of the project, the crew is very small. The field leader is able to keep up with the tasks that need to be completed, meet the schedule of the general contractor, and fill out the required documentation. As the job begins to pick up pace, the field leader is surprised by how difficult it is to keep a larger crew working efficiently. The field leader attends a lot of meetings with the field leaders of other trades and is continually pulled in multiple directions.

The field leader spends a lot of time rounding up tools and chasing down materials for the crew each day. It begins to seem like the crew is dictating what needs to be done and when, which causes production and morale to suffer. The field leader begins to lose ground and misses a few deadlines. The field leader decides the best course of action is to request an increase in the size of the crew. The project manager quickly declines the request because the project was bid very tightly and cannot absorb any additional labor hours. The project manager tells the field leader that the size of the crew is more than adequate and that there is no reason to miss the schedule.

The field leader is frustrated by the project manager's lack of empathy. The field leader puts in extra time each morning and afternoon and goes home exhausted every day. As the project falls farther and farther behind, the field leader begins to resent being put into this position without proper training. The field leader comes to the realization that, without some guidance and clear expectations, being successful is unlikely. The field leader decides to leave the company and go back to working as a crew member for another company.

Most construction field leaders began their careers as helpers, laborers, material handlers, or delivery drivers and then became an apprentice of the trade they were most interested in. Upon the completion of their apprenticeships, most of them chose to take the exam to become journeymen. After working as a journeyman for a period of time and exhibiting an aptitude for leadership, a company may offer a journeyman a field leader position. Having a strong work ethic and being technically sound at a trade are a great base for field leaders, but truly understanding what it takes to manage both the work and the people is much more complicated than simply reading prints, ordering materials, and building installations. Some field leaders make the transition from being a crew member easily, while others struggle with their new responsibilities. With experience and initiative, a field leader may eventually become a project manager, a general foreman, a superintendent, or even the owner of a construction contracting company.

Field leaders are almost always promoted from within the ranks of the company they work for, and every contractor is always in need of effective field leaders. The key to receiving advancement opportunities is to be dedicated and exhibit leadership qualities such as professionalism, respect, credibility, character, and integrity. Effective field leaders also use ethics when making decisions and take the time to teach others and learn more about their trade.

Project Planning

Project planning is one of the most important steps in the construction process. To be good planners, field leaders must have a solid working knowledge of the trade and the ability to visualize and conceptualize each phase of a project by studying prints and specifications. **See Figure 1-5.** In addition, they must have a good understanding of what each phase of the job will take in terms of manpower. The only way to keep a job running at maximum efficiency is to have a crew that is the correct size for the amount of work that needs to be done. A crew that is too large will burn valuable labor hours, and a crew that is too small will have trouble meeting schedules.

Planning Projects

Figure 1-5. To be good planners, field leaders must have a solid working knowledge of the trade and the ability to visualize and conceptualize each phase of a project by studying prints and specifications.

While project plans always change, a field leader's goal is to aim for the best solutions during each phase of the building process. An effective field leader also has contingencies built into the plan so that the crew will remain as productive as possible no matter the circumstances.

Before a construction project begins, the field leader must make sure to become as familiar as possible with the construction documents. This requires reading every specification and looking at every drawing. This is a very time-consuming task but one that pays for itself in the end. Effective field leaders also take the time to at least glance at the drawings for the other trades that they will be coordinating with throughout a project. During this first pass through the construction documents, it is important that the field leader focus on the overall concept of the project. The more focused planning and detailing comes later.

Taking a second pass through the construction documents is when a field leader begins making thorough notes about anything that may become "pinch points" for the crew. One example might be a special type of material that will be required for a later part of the project. An effective field leader will make a note of this so that the special material will be on site when needed. Discrepancies and typos in construction documents will also need to be clarified by an architect or engineer. Taking thorough and detailed notes of these items is crucial to a field leader's success. As a field leader spends more and more time working with the prints, the familiarity makes it much easier to find information when it is needed.

Once any large and time-sensitive issues have been resolved, the field leader begins breaking down larger tasks into smaller tasks and putting them in a sequenced order based on priority. For example, for the field leader of an electrical crew, a large task may be to mount a panel, run all the conduits (feeder and branch), pull wire, and terminate. The list of sequenced and prioritized tasks would look something like this:

- Mount panel E21C1B (1 person – 4 hours)
- Run 2 in. EMT from switchgear E2MDP (2 people – 3 days)
- Run (6) ¾ in. branch conduits to home run boxes (1 person – 3 days)
- Pull wire for panel E21C1B from switchgear E2MDP (4 people – 1 day)
- Pull branch wire for panel E21C1B (2 people – 1 to 2 days)
- Terminate panel and MDP (1 person – 1 to 2 days)

Many tasks must be performed before other tasks can be completed. For example, a panel cannot be terminated unless it has been mounted, the conduit has been run, and the wire has been pulled.

Another item that a field leader must take into consideration is how the work of another trade may affect what needs to be done and when. For instance, if an electrical field leader wants a crew member to mount panel E21C1B, but the wall that it is to be mounted on has not been built, then there is nothing that can be done. But if the list of tasks was put together early enough, the electrical field leader could ask the framing field leader to have the wall built by a certain date.

Another item that a field leader must take into consideration is how the work of another trade may affect what needs to be done and when.

Another advantage to breaking down large tasks into smaller ones is that it allows the field leader to carefully look at each line item and determine which tools and materials are required for each task. When reviewing a broad range of tasks without breaking them down into much more manageable pieces, it is difficult to pick up on the intricacies and special requirements that may need attention.

Crew Supervision

Field leaders are responsible for seeing that crew members do their jobs skillfully and efficiently. They must also be sure that assigned work progresses on schedule and reflects the quality standards of the company. When a field leader makes the transition from being someone who goes to work as a journeyman to someone who leads a crew, the perspective has to shift to staying 10 steps ahead of the crew at all times. An effective field leader tells the crew what is expected of them. If the field leader leaves too many questions unanswered or does not provide the crew with the necessary information, tools, and materials they need to do their jobs effectively, the project is guaranteed to suffer.

Leading By Example

Field leaders must also monitor the progress of the crew to reduce the possibility that tasks will not be finished on schedule. Effective field leaders comment on the overall quality of the work performed by their crew members. Through encouragement and reinforcement, crew members will continue to become more efficient and productive. Another responsibility field leaders have is to ensure that experienced journeymen are taking a reasonable amount of time to train apprentices. This long-standing tradition is something that is invaluable to not only the company but to the construction industry as a whole.

Leading in the Field: A Group Approach to Field Leadership

An electrical contractor is awarded a project to build out a warehouse space. The project was bid very tightly. It will take four months to complete and will require a crew of three journeymen and one apprentice. Every journeyman the company has available has been a successful field leader for them in the past. So rather than choosing one field leader, they ask all three individuals to combine their leadership skills to manage the project.

The three journeymen decide that they will each take on a different scope of work and try to stay out of the way of what the others are doing. They order some materials to get started, rent a scissor lift, and have the shop deliver a gang box with the tools they think will be needed.

From the very beginning, the crew has trouble getting into a rhythm. When one person needs the scissor lift, another person is using it. When one journeyman orders materials, one of the other journeymen takes some of it, leaving the first journeyman short. There is only one set of construction prints. So when one field leader is using them, another field leader has to spend valuable time walking the large site to find them. The general contractor becomes frustrated because none of the journeymen have stepped forward to coordinate with the other trades.

After four months of infighting among the three journeymen, the job is completed at a 10% overage of labor hours and a 12% overage of materials. The company managers hold a meeting to discuss how a project could lose so much money with three very capable journeymen-field leaders. The journeyman-field leaders say they only felt responsible for a third of the project and were not concerned about what the others were doing. Each also comments how they would rather have been a crew member for either of the other two journeymen-field leaders. All three say the project would have been a success if there was one clear leader that managed all the tasks, ordered all the materials, coordinated with the other field leaders, and was the single point of contact for the general contractor.

Field leaders are also responsible for communicating the organizational and production expectations to everyone on the crew. Without clear direction from a field leader, job sites become disorganized, which ultimately brings down efficiency, production, and crew morale. Effective field leaders define and communicate the role that each person on the team has in maintaining an organized job site. Most of the responsibility for job site organization often falls on the shoulders of the material handlers and laborers. Their willingness to have the tools and materials where they need to be, when they are needed, increases efficiency and production for the entire crew. Material handlers and laborers who feel respected and are treated as valuable members of the team are much more likely to maintain an organized site. An effective field leader who provides encouragement and shares knowledge to help these entry-level workers enter a trade apprenticeship and begin to reach their career goals will have committed team members in the short term and allies for years to come.

Project Coordination

Field leaders are responsible for coordinating activities on the job site to make sure that tasks are completed safely, according to code, on time, and within budget. Field leaders are also responsible for solving any problems that arise during production and documenting any changes to the prints. In addition to problem solving actual installations, effective field leaders also address disruptive behaviors that lead to conflicts among people. This may involve mediating a conflict between two crew members or finding a way to get along with the field leader of another trade.

Effective field leaders deal with individuals who do not do what they are asked to do and constantly coordinate with field leaders of other trades. Ignoring poor performance by a crew member and failing to coordinate with another trade only increases the likelihood that a field leader will have rework to complete.

Effective field leaders not only coordinate all project activities but also problem solve to get the job done on time and within budget while meeting customer expectations. Without an effective field leader serving as the principal planner, organizer, safety leader, decision maker, problem solver, documentarian, communicator, and teacher, a construction project is often not successful.

Section 1.3 For Discussion

1. What makes a person qualified to be a field leader?
2. Describe the essential role of project planning in field leadership.
3. How does a field leader interact with a crew?
4. Describe how a field leader coordinates a project.

Section 1.3 Take Action

1. Draw a career path listing the jobs leading up to becoming a field leader.
2. Role-play a field leader interacting with a material handler who wants to become an apprentice.

The Contracting Business

Case Study

Steve and Paul have worked well together as journeymen for the past 10 years. They decide that their combined experience would make them great business partners. So they open a new electrical contracting company, S & P Electric. Steve manages the employees, prepares estimates, and serves as project manager. Paul is in charge of finances and marketing. He also makes sure customer needs are being met.

Steve and Paul each take out equity loans against their homes to put together the initial working capital needed to start the company. They also open a line of credit in order to lease a building and office equipment; build out the space; and buy office furniture.

Part of their business plan is to take advantage of the relationships they have acquired over the years while working for other companies. Before opening S & P Electric, Steve and Paul were able to secure two small projects that allowed them to hire a few employees and generate some revenue. Their long-term objective is to bid on larger projects to make money.

March

During the first month, Paul spends most of his time marketing S & P Electric to local general contractors and other potential customers. His primary objective is to inform potential clients within their market area that there is a new contractor in town that provides quality electrical installations and excellent customer service. He also sits down with material suppliers to negotiate the discounts on larger orders and determine what the payment terms are.

Steve spends his days supporting the new employees. He stops by both small projects every day to keep an eye on progress and make sure things are running efficiently. He also makes sure that the customers are satisfied with the progress being made. When Steve has time at the office, he writes purchase orders for new tools and gang boxes so they will be available when S & P Electric wins a larger project.

At the end of the month, Steve and Paul sit down to discuss how things have gone and their future plans. Paul informs Steve that they will receive a couple of bid opportunities at the beginning of April. Steve tells Paul how things are going on their two current jobs. They both agree that the company is heading in a positive direction.

April – Weeks 1 and 2

During the first couple weeks, S & P Electric receives two requests for proposals. With tight submission deadlines, Steve is forced to work long hours to complete them on time. One proposal is for a tenant-improvement project in a downtown office building, and the other is for a new fast food restaurant being built from the ground up. During this busy week, Steve decides to have one of his new employees stand in at a coordination meeting with the customer of one of the current jobs.

Meanwhile, Paul looks over the company's finances and realizes that the previous month's payroll took a larger chunk out of their working capital than anticipated. He begins writing a business forecast to present to their loan officer to ask for an increased line of credit if they land one or both of the new projects.

Case Study ...continued...

April – Week 3

Steve receives word that they won the restaurant project and that it is scheduled to begin in a couple of weeks. Steve and Paul are ecstatic to land their first competitive project. Paul sends out the billing for all of the work they completed and wages paid during the month of March.

April – Week 4

Paul receives a couple of large invoices from tool suppliers and asks Steve to look them over.

Paul:	"Steve, take a look at these invoices. There has to be a mistake."
Steve:	"No. No, these are right. I ordered and received all these items."
Paul:	"I don't understand! You're telling me that we needed all those tools and gang boxes for those two small jobs?!"
Steve:	"Well... no. I was planning ahead for when we landed a larger job, which we just got."
Paul:	"Look, we don't have the money for all these things right now. You're going to have to send it all back. We can buy these things when we absolutely need them."
Steve:	"We can't send it all back. I already had one of the apprentices paint our company name on them."
Paul:	"I would think you'd know better! We have payroll to make, and I wasn't planning on having to pay such a huge bill this month. Next time, talk to me before you do something like that!"
Steve:	"You need to relax, Paul. You're going to give yourself an ulcer. Everything's going to be fine."

May – Week 1

Paul receives a call from a customer of one of the small jobs. The customer is dissatisfied with the employee sent to the coordination meeting. The employee is disorganized, unprepared, and argumentative. The customer also says that the only reason he gave S & P Electric the project in the first place was because he trusted both Paul and Steve. He expected one of them would manage the project, not an unknown employee.

Paul apologizes and promises that either he or Steve will be at all future coordination meetings.

Case Study ...continued...

May – Week 2

Steve hires two more journeymen electricians and one apprentice. He assigns one journeyman as the field leader on the restaurant project, and he assigns the other journeyman and apprentice to be his crew.

The journeyman assigned as the field leader is one of Steve's former apprentices. Steve knows the young journeyman is a hard worker and hopes that he will be an effective field leader.

May – Week 4

Paul receives payments for the bills sent out in April. These bills covered the expenses incurred on the two small projects during March. For a moment, he feels a bit of relief, but it does not last. The extra manpower they hired for the new project is putting an extra squeeze on their finances.

Steve and Paul sit to discuss business and make sure they are on the same page.

Steve:	"Hey, did we get those checks?"
Paul:	"Yes, we got them earlier this week."
Steve:	"Great! Let's buy some work trucks. We've been working hard. I say we treat ourselves! I've had my eye on the new crew cab that just came out last fall."
Paul:	"I really hope you're trying to be funny."
Steve:	"What's the matter?"
Paul:	"I have a meeting scheduled with the bank first thing next week. We need to talk about increasing our credit line. We're close to not making payroll.
Steve:	"How can that possibly be?! After all the money we put into this thing?!"
Paul:	"Remember those tools and gang boxes you bought? That was enough money to cover our entire payroll for three weeks."
Steve:	"Seriously?!"
Paul:	"Anyway, I'm trying to take care of that, but there's another issue we need to talk about."
Steve:	"Great. What?"
Paul:	"I know you're busy with the new project, but I just got another call from one of our customers. You know the one that helped us get started?"

Case Study ...continued...

May – Week 4

Steve:	"Yeah?"
Paul:	"Well, he's not happy. He said you missed another coordination meeting. We've talked about this. You can't expect him to want to work with us in the future if we aren't fulfilling our commitments to him."
Steve:	"I can't be in two places at once! This new project has me running in a million different directions! Why can't that journeyman we hired figure all this out? He's getting paid a lot of money."
Paul:	"Agreed. But you and I are the ones that made the promises that need to be kept. You know, the last time I stopped by to see how things were going over there, I saw a guy from Smith Electric talking to the facilities manager. You know they would just love to push us out and take over the work."
Steve:	"Alright, alright. I'll be at the next meeting."

The restaurant project was completed on schedule, and S & P Electric made enough money to cover their expenses and make a small profit. However, Steve decided to let his field leader go. Although he was a hard worker, Steve found out the hard way that planning, organizing, and leading people was not his strength.

In the summer, S & P Electric won two new projects that kept them busy through the end of the year. They continued to struggle to make ends meet.

TAKE A CLOSER LOOK

1. How does managing a contracting business differ from managing construction projects as a field leader?
2. How does this case study highlight why contractors need to employ effective field leaders?

Case Study ...continued...

How Can a Contractor Increase Potential for Success?

When bidding work:

- Take a thorough approach when estimating projects.
- Find creative ways to bid at the lowest possible price.
- Purchase what is needed and not what would be "nice to have."

When selecting manpower:

- Invest in the people employed and make sure they are properly trained in field leadership.
- Select field leaders who have some tenure with the company and some experience being successful.
- Have project managers mentor potential field leaders.

When considering finances:

- Make sure field leaders are aware of the financial implications of every decision made on the job site.
- Keep lenders informed of when financial needs fluctuate.

When providing customer service:

- Stand out from competitors by providing value and quality installations without cutting corners.
- Share the business vision with the field leaders and explain how they play a part in that vision.
- Monitor customers' perceptions of the quality of work being done.

PERSONAL PERSPECTIVE

1. Describe a situation in which a contractor won a project based in part on the field leader(s) hired.

Case Study ...continued

PERSONAL PERSPECTIVE

2. Describe a situation in which a contractor selected a field leader who did not have the leadership skills required for the job.

3. Describe a situation in which a field leader did not consider financial implications as project decisions were made.

4. Describe a situation in which a field leader did not make sure the customer was pleased with the project's progress or quality.

chapter 2

LEADERSHIP QUALITIES

The fact that a person has been asked to be a field leader recognizes years of hard work and means that the individual is viewed to have leadership potential. However, going from being a reliable crew member to being a person who is constantly in charge is not an easy shift to make. A new field leader must move away from thinking only of the actions required for an individual to be productive on the job to anticipating the needs of all the people involved in a project from start to finish.

The role of a field leader comes with new responsibilities and added stress. An attitude of constant anticipation becomes the new norm. A field leader must work hard every day and face many challenges. Along with the many years of experience required, there are eight key leadership qualities that enable a field leader to be successful. Effective field leaders consistently practice professionalism and respect, have proven credibility and character, exhibit ethics and integrity, and teach future leaders as well as learn from others.

Section 2.1: Professionalism and Respect

- Describe how professionalism is demonstrated by effective field leaders.
- Explain why effective field leaders need to respect crew members, field leaders of other trades, and members of the construction management team.

Section 2.2: Credibility and Character

- Describe how effective field leaders demonstrate credibility.
- Explain why effective field leaders need to exhibit character.

Section 2.3: Ethics and Integrity

- Describe how effective field leaders use ethics to make fair decisions.
- Explain why effective field leaders need to exhibit integrity.

Section 2.4: Teaching and Learning

- Describe how effective field leaders teach and mentor crew members.
- Explain why it is important for effective field leaders to continue to learn.

Learner Resources
ATPeResources.com/QuickLinks
Access Code: **335421**

Section 2.1

PROFESSIONALISM AND RESPECT

Contractors count on field leaders to not only be competent at their trade but to set an example by behaving in a professional and respectful manner. These two leadership qualities often make potential field leaders stand out among other crew members when they are apprentices and journeymen. Anyone can learn what it takes to be a field leader, but not everyone can apply that knowledge with the focus and commitment of a true professional.

> Professionalism requires being courteous, honest, and responsible when dealing with individuals or companies.

Professionalism requires being courteous, honest, and responsible when dealing with individuals or companies. Effective field leaders are always professional in their interactions with crew members, with field leaders from other trades, and with members of the construction management team.

Effective field leaders are also respectful representatives of the contractor and the customer. Construction workers are a diverse group of individuals who come from many walks of life. Field leaders work with people who do not share the same hobbies, political views, religious beliefs, or philosophies. Effective field leaders respect others while remaining focused on completing projects on time and within budget.

Leading in the Field: Shifting Mindset

On the first day, a new field leader may not be quite sure where to start and choose to help the crew. Within a week, a new field leader often becomes inundated with questions from the general contractor, other trades, and the project manager regarding scheduling. At different times, the crew seems to run out of materials, causing them to have to shift tasks while waiting for more materials to arrive. A new field leader may wonder, "How am I supposed to get things done when all these issues keep cropping up?" A field leader who does not spend time planning and organizing sets up the crew for wasted production time.

A new field leader soon realizes that having worked for other field leaders was not enough to prepare for this new role. Transitioning from working as part of a crew to managing a crew not only requires a thorough understanding of the tasks required but the ability to build a comprehensive plan for accomplishing those tasks within a specified timeframe and budget. Becoming an effective field leader takes time, experience, and determination.

The success of an effective field leader is determined by how project setbacks and daily encounters with people are handled. It takes intestinal fortitude to face challenges each day on a construction project and remain professional. For example, if a crew member reported that an installation cannot be completed based on the information provided, a field leader may become annoyed or jump to conclusions. However, an effective field leader would revisit the installation, try to figure out why it cannot be done the way it is laid out, and come up with a viable solution for moving forward.

Field leaders interact with many different people each day and must maintain a professional attitude during each personal interaction. No matter whom a field leader is talking to and no matter what position they hold, it is always best to treat each person professionally in order to earn their respect. For example, an effective field leader would patiently listen to customer needs, even when an alteration to the building plans requires change orders and changes to the overall schedule and project plan.

Professionalism

Set the Example

Effective field leaders set a professional example by wearing personal protective equipment and work-appropriate attire; by having a positive attitude about safety and job-site policies; and by communicating their attitude and work ethic through words and actions. The professional attitude of an effective field leader influences the attitude of the crew, which increases morale and productivity. Field leaders who do not take safety and job site policies seriously set a poor example for crew members.

An effective field leader also sets a professional example by providing crew members with appropriate tools and materials and by providing the support required to accomplish the tasks laid out for them. A crew who witnesses the effort a field leader brings to a project will work hard to meet expectations. Think about the alternative. If a crew of 10 people were left to find something to work on and to purchase materials and supply their own tools, how profitable would the project be? What if that crew only had one set of prints and had to take turns trying to figure out what they were going to do? The project would definitely not run efficiently, let alone make a profit.

Leading in the Field: Missing Materials

A field leader lays out a task for a crew member. The crew member asks if the materials are on site. The field leader says, "I bought some a while ago… I'm sure they're here somewhere. Take a look around. Let me know if you don't find them."

What if the material was not on site? How productive would it be for the crew member to have to look all over the job site to find materials? What about all the time wasted? Any time a crew member has to figure out something that the field leader should have provided, the job is losing money. A crew is only as productive as they are led to be. A field leader is responsible for providing all the tools, materials, and information the crew is going to need.

In a competitive marketplace where multiple companies provide the same services for around the same price, the level of professionalism exhibited can tip the scale. Many opportunities are missed because a potential customer has a less-than-positive perception of how projects are being managed. An effective field leader sets a professional example by giving customers the opportunity to see a job site that is a cut above the rest. Sprucing up a job site for an important visitor is not the way to go about it. Rather, it is about having pride in the organization of the job site and making sure that it runs efficiently. Staying organized also lessens the stress levels of everyone involved.

Leading in the Field: What Visitors Perceive

The job superintendent and the customer arrive on the job site unannounced. The customer says, "We were in the neighborhood and decided to stop in." The job trailer is a disaster zone, with piles of paper and wrinkled prints that have been taped together spilling onto the floor.

Visitors make judgments about leadership abilities based on what they see and hear, including obvious signs of disorganization. Subconscious or not, these judgments affect how a field leader is perceived. If visitors see an organized trailer on a job site with adequate tools and materials and a crew that takes pride in their work, they will perceive the field leader to have high standards that will result in a quality product.

Respect Is Earned

In the construction industry, the word "respect" is used frequently. "Respect journeymen." "Respect other trades." "Respect management." Yes, individuals in positions of authority and those who have been in the business for a long period of time deserve a certain level of respect. However, rank alone does not automatically command respect. Respect is earned by being worthy of it. An effective field leader is someone the crew can count on, someone who makes sure customers get what they want, and someone who helps everyone make a steady income. An effective field leader earns the respect of the crew, other field leaders, and members of the construction management team.

Respect is a social contract between two people. Mutual respect sets the foundation for positive relationships. When one person is unwilling to participate, it destroys any opportunity for the relationship to succeed. And since every person in the construction business could potentially cross paths, it is important to remember that someone who is a crew member today could be a field leader in the future.

Field leaders who do not respect others cannot expect to receive respect in return. **See Figure 2-1.** Effective field leaders respect others as well as themselves. There are very few, if any, individuals who find their way into a field leader position without first putting in their time in the trenches. There is usually an apprenticeship that may take as many as five years to complete.

The experiences an apprentice or journeyman has working under field leaders can negatively impact how authority is viewed. For example, a new field leader who worked under field leaders who browbeat crew members may learn how to be unprofessional and disrespectful. Likewise, a new field leader who worked under field leaders who ruled by intimidation often learns that behavior. Authority should only be used to execute a successful project, not to hold other people down. Effective field leaders create a positive working environment where every person is treated respectfully.

Figure 2-1. Field leaders who do not respect others cannot expect to receive respect in return.

People perceive others based on what they have heard, seen, or experienced. Perceptions can enhance or inhibit respect. Sometimes perceptions are a fair assessment, and other times they are far from accurate. When a field leader is basing interactions upon perception of a crew member or a project manager, those same individuals are also basing their interactions on perceptions at the same time. For example, a crew may start a job with a less-than-flattering perception of a field leader because one crew member may have told them that the field leader was untrustworthy and unorganized on other jobs. True or not, this shared perception will cause the field leader to have to win the crew's trust on the current project.

While the importance of a good first impression should not be overlooked, it is the way a person is perceived over time that matters most. For example, if a field leader takes offense at the way a crew member looks, it is a short step to making an assumption that the person has no work ethic. If observations are coupled with comments about an individual, a lesser opinion is formed that is both unfair and disrespectful.

Leading in the Field: Demanding Respect

What if a field leader takes the attitude, "If people don't like the way I do things, they can hit the bricks." While it is human nature to want to be liked and respected, the "my way or the highway" approach does not work, has never worked, and will never work. Field leaders who operate from a superior perspective breed poor morale, poor communication, poor production, and ultimately poor profitability. Basically, it comes down to this – every field leader must learn how to work with people and show evidence of leadership by example. If not, they face a long uphill battle that will thwart success every step of the way.

There is no way to know what perceptions people have of one another. That is why it is important to understand that every field leader is viewed differently by the perception of each individual. If one person on the crew does not like a field leader, then the rest of the crew may find reasons to feel the same way. Likewise, if a crew has to work for a field leader who is driven by ego and power, they will not show that field leader respect. Field leaders cannot allow insecurities or the need to feel validated compromise effectiveness. Effective field leaders put personal wants and needs aside and build positive working relationships based on mutual respect. Over time, effective field leaders use planning, organizational, and people skills to earn respect.

Section 2.1 For Discussion

1. Explain how effective field leaders exhibit professionalism.
2. How can effective field leaders demonstrate respect for crew members?
3. Explain how effective field leaders demonstrate respect for other field leaders.

Section 2.1 Take Action

1. Share a past experience in which the professionalism exhibited by a field leader inspired a crew to perform at maximum productivity.
2. Share a past experience in which the respect a field leader showed a crew caused them to be more productive under a tight schedule.

Section 2.2

CREDIBILITY AND CHARACTER

Effective field leaders are known for their credibility and character. These two leadership qualities are what separate effective field leaders from average field leaders. Credible leaders make difficult decisions on a daily basis, even when the options are not always clear. These individuals are committed to their word and have the character to stand up for their principles. Credibility is a combination of competence and character. It is as simple as people believing what a person says based on the history of commitments being met and the intent of the individual's actions. Credibility is built or torn down with every personal interaction.

From the first day a field leader interacts with a crew member to each encounter with the field leader of another trade, the "credibility meter" goes up or down over time based on how actions match words. Actions are what determines which way the needle moves on the "credibility meter." **See Figure 2-2.** For example, a field leader coordinating with the field leader from another trade reports that the crew will be out of a particular area by a specified time. But if they have not finished when the other crew shows up to go to work, the field leader of the other trade has to redirect that crew to another part of the job, wasting time and production. Even if this is a one-time occurrence, it sets the "credibility meter" back a notch. Because of the incident, the field leader of the other trade will approach the next encounter with skepticism.

The "Credibility Meter"

Figure 2-2. Actions are what determines which way the needle moves on the "credibility meter."

A field leader can recover from a misstep now and then. However, credibility is lost when a field leader continually makes statements that are not backed up by actions. An effective field leader understands that being credible is just as important as technical skill.

Make Sound Decisions

Field leaders are responsible for making informed decisions regarding safety, efficiency, and production on the job. The only way for a field leader to manage this responsibility is to have a solid working knowledge of all safety procedures, code requirements, and job-specific construction documents. If a field leader chooses not to reference this information when determining how to approach a task, there is no way to verify that proper actions are being taken. For example, a crew member asks a field leader how to anchor the frame of a large motor to a concrete slab. Rather than take the time to reference the equipment manufacturer's engineered anchoring specifications, the field leader tells the crew member to use the anchors they have on site. In addition to losing credibility, what will the field leader say three months later when the motor vibrates loose and the customer demands an answer? Effective field leaders take the time to reference all pertinent information when making project decisions.

Leading in the Field: Actions Speak Louder than Words

A new field leader hesitates to speak up during coordination meetings. The field leaders from the other trades all have many more years of experience, so the new field leader keeps quiet and tries not to look stupid in front of them. When asked if the schedule can be met, the new field leader simply says, "Yes." After two weeks pass, it becomes obvious that the new field leader cannot meet the schedule. In frustration, the other field leaders ask the general contractor to meet with the new field leader to discuss the problem. The general contractor explains that not actively coordinating with the other trades is stripping away the new field leader's credibility each time a commitment is not followed through. The new field leader agrees to speak up in future coordination meetings and share why tasks can or cannot be done according to the planned project schedule.

Take Responsibility

Field leaders have an enormous amount of responsibility. A field leader is not just a member of the crew who puts in the hours and goes home. Field leaders are expected to do everything possible to ensure that a reasonable amount of profit is made from each project. Effective field leaders interpret prints to determine what is needed to run the most efficient and productive job possible while making sure the crew works safely. They are also coaches and mediators who bear criticism from project managers and field leaders of other trades on a project. Effective field leaders also have the character to be held accountable when crew members make mistakes.

Character is what makes up a person on the inside. A person of character does the right thing, even when no one is watching. Abraham Lincoln once said, "Character is like a tree and reputation like a shadow. The shadow is what we think of it; the tree is the real thing." Effective field leaders do not let obstacles stand in the way of what is right. Even when the burden is heavy, an effective field leader stays steadfast, sees challenges as having solutions, and makes the necessary adjustments to keep projects on track. It is the culmination of many events that build character or break it down.

A field leader who lacks the character to do the right thing for a customer or a crew will struggle to succeed. Field leaders may not think that people are paying attention to all the decisions being made. But whether they do it consciously or not, people subconsciously keep mental notes of the actions of those around them. Field leaders who hope to disguise their shortcomings for as long as possible eventually run out of opportunities to lead a crew. If given a choice, crew members will not choose to work for a field leader with no character. And a customer who feels deceived or disrespected by a field leader will not want that field leader to run a project.

Leading in the Field: Taking Responsibility Reflects Character

A crew member breaks an expensive tool. The field leader calls the project manager to report that the tool will need to be replaced. A few hours later, the project manager calls the field leader to share that the company owner is angry and wants the name of the person who broke the tool. The field leader replies, "If he wants a name, give him mine."

Why would a field leader respond like this? The field leader knew it had been an accident and made the decision not to put the crew member in a position of probable disciplinary action. An effective field leader takes responsibility for the actions of the crew. And crew members want to work for field leaders who place loyalty above the need to look good to management.

Effective field leaders have thick skin, broad shoulders, a can-do attitude, a self-motivating drive, and multitasking capabilities. Bringing anything less to the leadership role produces less-than-optimum results. No one likes to be in the line of fire when things go wrong. But if a field leader skirts the responsibility that comes with the position, it speaks to a lack of character that will not be respected on the job site.

Section 2.2 For Discussion

1. How do effective field leaders establish credibility on the job site?
2. Why is it important for effective field leaders to have character?

Section 2.2 Take Action

1. Share a past experience in which the credibility of a field leader was questioned.
2. Share a past experience in which a field leader exhibited character.

Section 2.3

ETHICS AND INTEGRITY

Being ethical involves making decisions based on fairness in order to determine the best outcome in any questionable situation. **See Figure 2-3.** There is usually more than one way to make a particular installation. But on almost every project, a team of architects, engineers, designers, and the customer have gone to great lengths to clearly spell out the construction materials and installation methods to be used. Sometimes these decisions are based on code requirements, and other times they are based on preference.

A field leader's ethics are tested as the pressure of managing timelines and meeting expected profitability increases, making shortcuts appear to be plausible options. This type of unethical choice opens the door for field leaders to make ill-advised decisions, such as choosing substandard products or taking shortcuts in an attempt to increase profitability. Unfortunately, some field leaders may bank on the fact that it is highly unlikely anyone is going to inspect work every single day.

Ethical Decisions

Figure 2-3. Being ethical involves making decisions based on fairness in order to determine the best outcome in any questionable situation.

Effective field leaders do not consider unethical options. They choose to manage projects in an efficient manner that fulfills the commitment the company made to the customer based on a signed contract.

Leading in the Field: Meeting Customer Needs

A field leader completes a couple of difficult projects while working inside an existing hospital. These projects were challenging because the areas of the hospital that were not being upgraded continued to operate while the construction crew worked. The field leader listened to the customer's concerns and adhered to every policy and request made, all with minimal disruption to the hospital and its function.

At the conclusion of these projects, the customer writes a letter to the contractor commending the field leader's attention to detail and concern for the special needs of the hospital. The letter closes with a request that the same field leader work on their future hospital projects.

Effective field leaders also exhibit integrity—a sense of what is right and a commitment to leading by example. When faced with situations where it would be easier to pass the buck, the field leader is ultimately responsible for the mistakes of the crew and the project as a whole. A crew who thinks their field leader will throw them under the bus when the pressure turns up will be less than forthcoming with information when mistakes are made. No one expects a field leader to be everywhere at once. However, an effective field leader stands behind the crew even when things go wrong.

Treat People Fairly

By treating people fairly, effective field leaders demonstrate how to work together toward common goals. Effective field leaders treat people fairly because it is the right thing to do. It is not always easy to be fair when tight schedules and profitability are in play. Effective field leaders know that every person pays close attention to how others are treated and subconsciously keeps score. For example, if it becomes apparent that a crew member is receiving preferential treatment, the rest of the crew will lose respect for the field leader, which often results in poor morale. Likewise, while bird-dogging a crew may give a field leader the illusion that everything is under control, crew members will view this action as intimidation. Effective field leaders know that having authority over crew members is not what motivates them. Consistent leadership that is ethical in both word and action sends the proper message.

Crews who work as a team are always more productive than crews who do not. However in every trade, there are tasks that are considered less desirable than other tasks. It is important that the same crew member is not always assigned the "worst tasks." All crew members need to feel that their contributions to the project are valuable. Effective field leaders treat crew members fairly and help them focus on completing projects on time and within budget.

Be Honest and Trustworthy

People want to believe in the people for whom they work. Honest and trustworthy field leaders earn the respect of crew members, field leaders of other trades, and management alike. Field leaders may have to answer difficult questions about a project or about the company running a project. Crew members see the field leader as the one they should be able to trust to address their concerns. They need to believe what the field leader says. As the adage goes, "Trust takes years to build, seconds to break, and forever to repair."

Leading in the Field: Mutual Trust

A field leader is managing four floors of a 20-story office building and has a few crew members on each floor. There is no way to spend enough time on every floor to spot all the coordination issues and material shortages or notice every print discrepancy. The crew members take it upon themselves to help the field leader by providing pertinent information that allows problems to be addressed before they cause a disruption in production. But how can the field leader be sure that crew members will be compelled to share this vital information? The answer is because the field leader is honest and trustworthy and treats the crew with respect.

A field leader who beats around the bush instead of giving people straight answers or dodges questions with ambiguous answers will be viewed as untrustworthy. If a crew thinks the field leader is not shooting straight, a mental mutiny will set in. The crew may even find satisfaction in doing things to "even the score." The last thing a field leader needs is to motivate a crew to waste time. For example, a crew of 10 can generate a high volume of wasted labor hours by simply slowing down for the sake of punishing a field leader. They may take long lunch breaks, fail to report that materials are getting low, or just pay less attention to the details of various tasks. It only takes a few minutes here and a few minutes there before an hour has been wasted without any production taking place. The pressure of making sure a project runs efficiently and productively makes it easy to jump to conclusions when it is perceived that things are not being done quickly enough. Effective field leaders are honest and trustworthy, removing the need for others to doubt their integrity.

Section 2.3 For Discussion

1. Explain how effective field leaders can make decisions that are fair to everyone.
2. Why is it important for effective field leaders to demonstrate integrity?

Section 2.3 Take Action

1. Share a past experience in which the ethics of a field leader inspired a crew or the field leader of another trade.
2. Share a past experience in which a field leader exhibited integrity.

Section 2.4
TEACHING AND LEARNING

Effective field leaders are constantly teaching and learning. Effective field leaders know that teaching potential field leaders is the best way to ensure a future workforce. After working as an apprentice and journeyman for years, a field leader has the trade experience to teach crew members how to complete tasks that are performed on a frequent basis. Effective field leaders acknowledge the training, education, and experience of others, and share that information with crew members as they direct day-to-day performance on the job. They make sure tasks are completed according to code, on time, and within budget. Field leaders also support crew members by solving problems that arise to ensure that production stays on schedule.

> Effective field leaders acknowledge the training, education, and experience of others, and share that information with crew members as they direct day-to-day performance on the job.

One of the greatest rewards of developing expertise is being able to share that knowledge with others. A journeyman working on a crew with other journeymen and apprentices has ample opportunity to teach others. While field leaders have fewer teaching opportunities because of the time required to plan and coordinate projects, they can listen carefully to crew members, answer their questions, and demonstrate how to perform a task when opportunities arise. They want to help potential field leaders grow into the job. Effective field leaders also know the critical importance of staying abreast of code changes, new installation methods, and technological advances in the construction industry.

Mentor Future Leaders

Effective field leaders often mentor, or coach and guide, a less experienced person on the job site. A mentor has a thorough understanding of how tasks need to be performed, can provide accurate information, and can teach another person how to perform tasks properly. **See Figure 2-4.** Proper instruction from a mentor reduces the chance that a crew member will need to be retrained. Mentors also provide constructive feedback; help others assess individual strengths and weaknesses; offer advice based on their experiences; and help others develop strategies to achieve career goals.

Every crew a field leader manages is composed of people with varying levels of expertise and career goals. Some crew members do not have any interest in becoming future field leaders. There are, however, crew members who want to be more than just crew members. These individuals may share their career goals, or an effective field leader may see potential in a crew member who does not recognize that potential.

Effective field leaders mentor aspiring crew members by explaining what is required to move to the next level and then helping those individuals acquire leadership skills. This can be accomplished in simple steps, such as showing a crew member how to detail a task or to order materials.

Teaching

Mentoring takes time but is well worth the effort. Some field leaders may also choose to inform the project manager that a crew member with leadership potential is being mentored.

Mentoring

Figure 2-4. A mentor has a thorough understanding of how tasks need to be performed, can provide accurate information, and can teach another person how to perform tasks properly.

Learn Continually

Some craftsmen believe there is no need to continue learning after obtaining journeyman status. While some trades do require continuing education classes to maintain a license, many classes focus on maintaining knowledge rather than learning about new materials and installation techniques. Updates to codes are an exception. Codes are continually being updated and changed. A code oversight could result in an illegal installation. And while the mistake is being repaired, scheduling conflicts with other trades could occur, not to mention a loss in both production and profitability. An effective field leader stays abreast of code changes and the effects those changes have on current and future projects. Effective field leaders are always learning and encouraging others to do the same.

Leading in the Field: Code Changes

An experienced field leader is managing a high-rise apartment building. The field leader is confident about this type of installation and brags about not needing to check the code. Midway through the project, an installation fails an electrical inspection. When the field leader reads the correction notice, there is a reference to a recent update to the code. The field leader realizes that not staying current with code updates just cost the company a large number of labor hours. Being an effective field leader requires continuing to learn every day.

Effective field leaders are also perpetual problem solvers. From the moment they begin reviewing prints, field leaders try to think of all of the things that will be needed for a project. However until a building actually rises from the ground, it is only a concept on a piece of paper or a computer screen. In addition to problem-solving actual installations, field leaders also problem-solve conflicts among people on the job site. It may be a conflict between two crew members that requires a mediator or determining how to get along with a field leader from another trade. Constantly solving problems requires effective field leaders to learn more efficient coordination methods as well as new ways to help people resolve conflicts so projects can be completed on time and within budget.

Section 2.4 For Discussion

1. Explain why it is important for field leaders to teach crew members how to perform tasks.
2. Why is it important for field leaders to mentor crew members with leadership potential?
3. Explain how a field leader who continues to learn brings added value to each project.

Section 2.4 Take Action

1. Share an experience in which a field leader taught a crew member how to perform a task.
2. Share an experience in which a field leader mentored a crew member whose career goal was to become a field leader.
3. Share how continuing to learn inspires better job performance.

Leadership Qualities

Case Study

Steven Daniels is a field leader who works for Newtown Framing and Drywall. Steven spent almost his entire apprenticeship with the company and is highly regarded as a hardworking and productive craftsman. He has managed a few projects with small crews but is currently in charge of a large crew working on a huge retail complex. Recently, schedules have been missed, which has caused the general contractor and project manager to warn him to stay on schedule or be replaced as field leader and demoted back to being a crew member. Steven's plan for moving forward is to drive the crew to succeed.

Tuesday **2:33 pm**

Steven approaches two crew members, named Karl and Anthony, who are framing offices at the back of a store. He notices that they have not made much progress from earlier in the day and immediately becomes agitated.

Steven: "What's going on back here!? You two should be wrapping this up soon, but it looks like you've barely started!"

Karl: "The sketch you gave us this morning is missing some key dimensions, and there aren't any studs back here like you said there were. We've been looking for a set of actual construction drawings, but it was a complete waste of time as usual."

Steven: "So much for you being the top-notch framers everyone told me about! You two are a couple of clowns! You know we have to turn this over to the electricians in the morning, right?! When were you planning on letting me know about this?! You know what, forget it! I'm going to get my tools and get this framed myself! If the two of you had any brains, you'd take notes. You're about to see a professional at work!"

Steven leaves the area to get his tools, and Karl and Anthony have a moment to talk.

Karl: "What did I tell you? The guy is out of his league. No material, bad layout, and we're the ones who don't know what we're doing!? I don't care how good of a craftsman he is! It doesn't automatically translate into being able to manage a crew."

Anthony: "I'm not putting up with this! I've been in this trade too long and seen his approach fail too many times. I'd rather go find work at another company than put up with a snot-nosed field leader who's ramrodding his way through a project!"

2:47 pm

On the way to grab his tools, Steven stops to check in with a couple of other crew members. He notices that neither one of them are wearing safety glasses and reminds them that Newtown Framing has a 100% safety glasses policy. He warns them not to violate the policy again, or there will be consequences. However, Steven fails to realize that he forgot to put on his safety glasses and has been walking the project without them all afternoon. This does not go unnoticed by the two crew members who stop working to discuss the irony of the situation after Steven leaves.

Case Study ...continued...

Tuesday 2:56 pm

Steven returns to the where Karl and Anthony are working and barks out orders.

Steven:	"Karl, I want you to help me snap some lines and shoot down the track! Anthony, go find us some studs. Hurry up!"
Anthony:	"I'll be happy to do that just as soon as you tell me where they are."
Steven:	"Just go next door and get some from Brad and Terry."

Anthony approaches Brad and Terry in the next building and asks them for some of the studs stocked there. They reply that if Steven said to take the studs, then they are fine with it. However, their supply of studs will run out soon if Anthony takes some. Brad and Terry comment on how they had to go to three different areas to get enough studs to complete the work they were assigned. Anthony contemplates looking elsewhere, but he decides not to come back empty-handed. So he loads a cart with studs and heads back. In frustration, he decides not to tell Steven that taking the studs will prevent Brad and Terry from completing their work.

Steven, Karl, and Anthony spend the rest of the day and two hours of overtime framing so the area can be turned over to the electricians in the morning.

Wednesday 7:15 am

Steven gathers the crew to talk about how they can be more productive and to discuss how the company's safety policies need to be taken seriously.

Steven:	"I don't know what the problem is with all of you, but we've missed a few schedules now, and my job is on the line because of it. I'm not going to take the wrap for anyone on this crew that isn't interested in putting forth the effort to succeed. Just yesterday, I had to stay late to help Karl and Anthony frame–"
Anthony:	"I've heard enough out of you! Why don't you just get my check?"
Steven:	"What did you say?"
Anthony:	"Just call the shop and tell them I quit! And I want my money!"
Steven:	"You're not serious."
Anthony:	"I'm very serious. And you know what?! Don't worry about calling the office! I'll just drive over there right now and tell them why I'm quitting!"

As Anthony leaves, Steven continues telling the crew how they need to be more productive. Steven also states that multiple crew members are ignoring safety policies and discipline will be enforced for anyone not wearing PPE. A crew member mentions that Steven was not wearing safety glasses on the job site the day before. The same crew member goes on to say that Steven can be frequently seen on the job site without proper safety gear. Steven ignores the comments and tells everyone to get back to work. As the crew heads back to work, Brad and Terry approach Steven.

Case Study ...continued...

Wednesday **7:20 am**

Brad:	"Steven, we don't have any studs left as of yesterday quitting time. We can't complete the area we were framing yesterday."
Steven:	"What?! Why didn't you guys say something before?"
Brad:	"We stopped by the trailer on the way out last night to let you know, but you weren't there. We left a note on top of your paperwork."
Steven:	"Anthony never said anything to me about it, and I didn't see your note. I knew he was trying to sink me! I'm glad he quit!"
Terry:	"I think Anthony leaving is bad for this project and for the company. I've worked with him for years. He's a good friend of mine. He's an asset to any field leader that keeps the layout and materials in front of him. I've never seen him get angry and walk off a job before. I really hope the company finds another project for him to work on."
Steven:	"You might want to reconsider who you hitch your wagon to, because that's not the impression I have of him. I really need you two to finish what you were working on. I'll call an order into the supply house and have you drive down there with the site truck and pick up some more studs. We can't miss another schedule!"

7:41 am

Steven heads to the trailer to try and get the detailing done for some of the upcoming areas the crew will be working in. The superintendent is waiting for Steven at the trailer and asks to walk the project to discuss some concerns. Steven takes the general contractor to the area that he, Karl, and Anthony framed the afternoon before. There are scraps of material lying around the floor, making it difficult for the electricians to safely get their work done.

Superintendent:	"It's great that you have this area framed on time, but the way this area was left is unacceptable. We've discussed this in our weekly coordination meetings. Each subcontractor is responsible for cleaning up after each task. It's not right that other trades either have to work around the mess your crew leaves or clean it up before they can work."
Steven:	"Between you and me, I've been having trouble with a few crew members who don't seem to take direction very well. In fact, I just got rid of one of the guys that left this mess. I'll remind my crew about picking up after themselves, and hopefully they'll get the idea that I mean business."
Superintendent:	"Okay. On to the next item. Why don't you have a crew working next door? You know it's on the schedule to be completed by the end of the day, right?"
Steven:	"It's another example of my crew not working with me. The two guys working in there ran out of materials without telling me, but I have materials on the way. I promise that even if we have to work overtime tonight, we will have that area done by tomorrow morning."
Superintendent:	"Okay. The other trades are counting on you."

Case Study ...continued...

Wednesday **9:45 am** The project manager calls and asks Steven to meet him at the job trailer.

Project Manager:	"I just came from the office where I had a long conversation with Anthony about what's going on. Based on what he told me, you've got some explaining to do. And just so we're clear about how the company views Anthony, we've convinced him to stay but we're moving him to another project. Now what's this I hear about you browbeating people in front of the whole crew?"
Steven:	"Seriously? That's what this is about? Did he tell you how I had to help him and Karl finish their work? I bet he left that part out!"
Project Manager:	"What he told me is that you are frequently running short on materials, and that the layout you are providing leaves a lot to be desired. Anthony also told me that you had them working overtime last night. You know that we didn't bid this project for any overtime. What was your thinking behind that? You can't just start working overtime whenever you feel like it!"
Steven:	"After the last few conversations with you and the general contractor about missing schedules, I didn't want to fall short again. I made the decision that I would rather get the job done than miss another deadline. I thought that was what you wanted me to do."
Project Manager:	"It's become clear to me and the other managers that you just aren't ready to manage a project this size. We're going to send out another field leader to take over the project to get it on track. You're very skilled, but you need to spend more time being mentored in leadership skills by someone with more experience before we can have you manage a large crew again."

TAKE A CLOSER LOOK

1. How did the crew respond to Steven's lack of professionalism?
2. Did Anthony overreact, and if so, why?
3. Can Steven expect the crew to follow safety policies if a good example is not provided?
4. How did the fact that Anthony quit instead of being fired for insubordination affect Steven's credibility as a field leader?

Case Study ...continued...

How Can a Field Leader Be Effective?

Demonstrate professionalism and respect by:

- Setting the example for others to follow
- Being courteous and responsible
- Respecting others
- Being competent

Demonstrate credibility and character by:

- Making sound decisions
- Being reliable
- Taking responsibility
- Being committed

Demonstrate ethics and integrity by:

- Treating people fairly
- Making ethical choices
- Being honest and trustworthy
- Standing by sound principles

Demonstrate teaching and learning by:

- Mentoring future field leaders
- Encouraging others to pursue career goals
- Learning continuously
- Taking construction leadership training courses

PERSONAL PERSPECTIVE

1. Describe a situation in which a field leader demonstrated professionalism.

Case Study ...continued...

PERSONAL PERSPECTIVE

2. Describe a situation in which a field leader demonstrated respect.

3. Describe a situation in which a field leader demonstrated credibility.

4. Describe a situation in which a field leader demonstrated character.

Case Study ...continued

PERSONAL PERSPECTIVE

5. Describe a situation in which a field leader demonstrated ethics.

6. Describe a situation in which a field leader demonstrated integrity.

7. Describe a situation in which a field leader demonstrated teaching (mentoring).

8. Describe a situation in which a field leader demonstrated learning.

chapter 3

PLANNING AND ORGANIZING

Contractors need field leaders who are willing to protect their interests as if they were their own. They need field leaders who can be trusted with their money and their reputation. A field leader who pays attention to detail by continuously planning and organizing will always have a job and be given other opportunities.

Construction plans have to be continuously reviewed and updated. Planning requires looking ahead as well as making adjustments based on daily occurrences. Learning to plan also includes learning to organize. A field leader who knows how to do both well will inspire crew members and motivate them to succeed. Furthermore, knowing what goes into thorough planning—in other words, what to do and when to do it—helps lessen everyone's stress and boosts their confidence.

Section 3.1: The Importance of Planning

- Explain how lists are the foundation of good planning.
- Identify eight potential consequences of failing to plan.

Section 3.2: Using a Planning Process

- Describe the four phases of the planning process.
- Complete a request for information (RFI).
- Produce a project plan using the four-phase planning process.

Section 3.3: Planning Challenges

- Describe common challenges to planning.
- Demonstrate how field leaders can maintain focus on the plan.
- Describe common interruptions and the best ways to minimize them.

Section 3.4: The Importance of Organization

- Identify potential staging areas for optimal organization.
- Explain how staying organized results in a safe work environment.

Learner Resources
ATPeResources.com/QuickLinks
Access Code: **335421**

Section 3.1

THE IMPORTANCE OF PLANNING

What exactly is planning? In brief, planning is the process of laying out all of the tasks, tools, materials, and manpower needed for a project and leaving as little as possible to chance or misinterpretation. The majority of field leaders believe it is enough to look through the prints a little every day and take note of some of the various tasks that will need to be addressed at some point. But frankly, this puts a field leader in a reactionary mode. If this is the status quo, there is no way the project will run efficiently. Each time a field leader has to react to an unplanned event, efficiency and production suffer.

Planning is the process of laying out all of the tasks, tools, materials, and manpower needed for a project and leaving as little as possible to chance or misinterpretation.

From the very first day of the project until the very last, an effective field leader never stops planning, weeks and even months ahead, for what is coming down the road and the best way to accomplish it. If a field leader is not planning every day, there will come a time during the project when this becomes apparent.

Yes, some things are hard to plan for. But those things are the exception, not the rule. If common emergencies come up during the course of a project, it is important to have a plan in place to handle them. For example, if temporary power and lighting may be required for a project, make sure to have the appropriate materials on site in case there is a sudden need for it. A field leader will never regret having planned. However, regret may arise over not having created a more detailed plan when something goes wrong.

Making Lists

Planning starts with making detailed lists of tasks that need to be done on a project. Jotting down a bunch of random ideas is not effective. Making a list is more than simply listing the elements of each task. Project lists include all the information (or layout), tools, materials, and manpower needed to ensure that specific tasks will be accomplished in the most efficient manner possible. **See Figure 3-1.** It is a waste of time to have a list of tasks to complete without having the tools and materials on site and readily available.

Project Lists

Run 2″ EMT from switchgear E2MDP (2 people – 3 days)
• 2″ EMT is on 2nd floor pipe rack
• 2″ EMT fittings and straps are on 2nd floor material rack
• Strut and threaded rod are on 2nd floor pipe rack.
• Hardware and anchors are fully stocked on 2nd floor material rack
• Roto hammer, bits, and extension cords are in 2nd floor gang box

Figure 3-1. Project lists include the information (or layout), tools, materials, and manpower needed to ensure that specific tasks will be accomplished in the most efficient manner possible.

Project prints are the starting point for making lists. By studying the prints every day and making notes of various tasks that will need to be addressed during the project, the field leader begins to create a project plan. It is important to note that the specific project plan that a field leader develops will likely differ from the base plan that may have been developed and presented by a project manager or an estimation team. Although the base plan provides a general overview regarding how the project should be managed and also serves as a starting point, it is essential that field leaders keep an open mind when detailing the specific tasks needed to complete the project.

An effective field leader knows that a realistic and workable plan can only be created by taking into account actual job-site conditions and required interactions with other trades. A project plan that does not coordinate task sequencing with other trades simply will not provide the continuity that promotes efficiency.

Planning and Organizing

Every moment of preplanning provides an opportunity to expose potential problems and find solutions before those problems occur. As soon as possible, an effective field leader gets questions answered and items ordered. When issues, additional questions, or long-lead tasks come up while looking through prints or during coordination meetings, the field leader adds these items to the list.

Leading in the Field: A Self-Imposed Crisis

A field leader did not take the time to create a project plan, so the project is not running very smoothly. Materials often run out, and the crew has to wait for deliveries to continue working. The crews of other trades are getting in the way, and other field leaders are finger-pointing over drops in production. Then a crew member calls in sick. The field leader does not know what to do.

A major task is scheduled for completion, and the crew runs out of materials—again! The field leader orders materials, but the supply house cannot deliver them immediately. So the field leader calls a taxi service. The taxi driver goes to the supply house to pick up the materials and then delivers the materials to the job site. The crew is working again. But now the field leader is wondering how to explain to the contractor why a taxi service was necessary.

This poor solution to a self-imposed crisis resulted from a lack of planning. The most basic responsibility of a field leader is to run an efficient project, and this cannot be done without a solid project plan.

Continuous, ongoing planning is necessary for long-lead tasks and items that require coordination with other trades. Even for shorter projects, planning two-to-three weeks ahead is necessary and should include tasks that might take as little as an hour or as much as a week or more to complete. Try to avoid getting caught up in day-to-day issues so that later phases of the project are not forgotten.

Using Lists

Field leaders are constantly bombarded with questions and concerns, and it is easy to forget something if it is not written down or added into project management software. A field leader who does not keep lists updated will get blindsided by problems that could have otherwise been avoided.

Not only is it necessary to have a system for staying on top of all of the things that come up throughout each day, but it is also necessary to decide how to take action on all the things added to continuously evolving lists. An organized list is better than a disorganized list every time. **See Figure 3-2.** The important thing to remember is that a list that stays in a pocket is nothing more than a piece of paper with scratches on it. After a day or so, it will get mixed with other loose papers and probably get lost altogether. And if it is ever found, the time to do what is necessary will have passed. Operating in a reactionary mode is one example of how a project can lose valuable labor hours. It also reveals why many field leaders believe they are hit with so many surprises.

Organized vs. Disorganized Lists

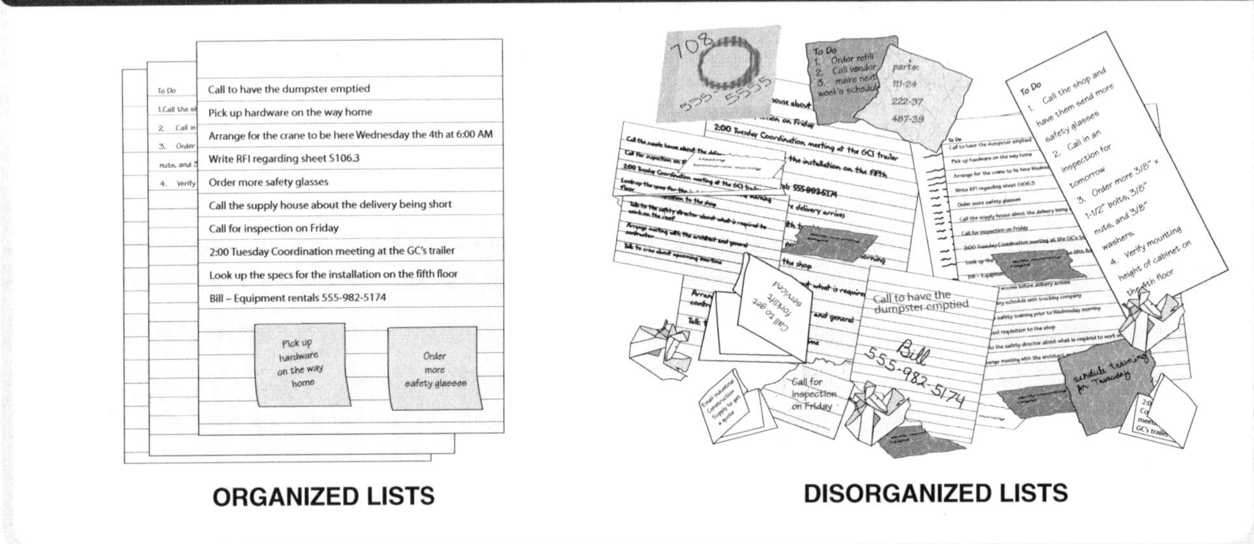

Figure 3-2. An organized list is better than a disorganized list every time.

Keeping ongoing lists is the foundation of good planning. The key is to form an educated plan for approaching any problems that may occur. That way, surprises can be minimized, and a peace of mind comes with knowing that there is a contingency plan. A field leader with an effective plan will run a project smoothly and will be confident that the challenges can be handled along with modifications that may occur throughout a project.

Failure to Plan

It is impossible to foresee everything that can or will happen on a project. But it is always within a field leader's power to have a thorough and deliberate plan. Preventing the issues that the field leader sees coming makes it easier to tackle the issues that cannot be foreseen. Is the crew standing around waiting for the field leader to decide what they should do next? Is it because of another trade or because they do not have tools or materials? Is it because not enough work was planned out in advance for the size of the crew? Whatever the reason, a field leader's failure to plan has consequences. Following are the likely consequences of failing to plan and how they can directly affect a project.

Loss of Profit. Although not every project can or will make a profit, the potential loss or gain of profit is always directly related to how well a project is planned and managed. On projects that do not have much of a chance to make a profit, an effective field leader can still minimize losses by planning well and managing an efficient crew. This approach is just as valuable to a contractor as turning a large profit on a project that has larger profit margins. When taking on a project that has obvious problems, remember that the project still needs to be completed. The better a project is planned and managed, the better it will be for everyone.

Unproductive Downtime. When a field leader changes direction because of another trade, it is easy to take the attitude that a plan cannot be followed because of them. If this is truly the case, then this is something the field leader needs to take up with the field leader of the other trade. But it still does not justify having nothing for crew members to do at any given time. Once labor hours are wasted, they are gone for good. **See Figure 3-3.** Having a contingency plan in place is what enables a field leader to handle these situations and not lose valuable labor hours. Furthermore, other trades will rarely be held responsible for slowing a crew down.

Wasted Labor Hours

Figure 3-3. Once labor hours are wasted, they are gone for good.

There are going to be times when what was planned does not turn out as expected due to unforeseen problems. This is simply the nature of the job. This is also the perfect time to take out that list of short projects. Remember, any production is better than none at all. It is the field leader's responsibility to make sure that there is always something productive for the crew to do.

Poor Crew Morale. The morale of a crew is critical to a project's success. Poor morale can be devastating. A common cause of poor crew morale is not having everything needed for a job while still being held to production expectations. Therefore, it only stands to reason that good morale can be maintained by supplying the necessary tools, materials, and information for the project. Contrary to what some may believe, most people show up to work wanting to be productive and do a good job. When crew members have what they need, they tend to work efficiently. Take that opportunity away, and the crew will quickly lose interest, and things will go downhill from there.

Unnecessary Rework. If a field leader pays attention to planning details, coordinates with other trades, and communicates information thoroughly to the crew, then a minimum amount of rework should be necessary. Projects never include rework hours—those hours go directly into the red column. It therefore makes sense that most rework comes from not paying attention to details when work is being planned.

One aspect of project planning is looking at all the prints from front to back and considering what the other trades will need to do their jobs. After a detailed idea is formed of the work required, the field leader must clearly communicate it to the crew and leave no room for misinterpretation. How does a field leader know when the crew is on the same page? A field leader not only needs to clearly explain a task but must also check in with the crew on an ongoing basis to confirm that they understand what needs to be done. Having crew members repeat instructions to clarify what they have heard is one way to catch mistakes before they happen.

On the job site, it is often important to take the time to provide a quick sketch to describe a task, including dimensions, materials, and any other pertinent information, to delineate what words may not. A sketch makes a good reference tool for a crew member who may forget a detail. **See Figure 3-4.** Time spent on this tactic is well worth the extra effort.

Reference Sketches

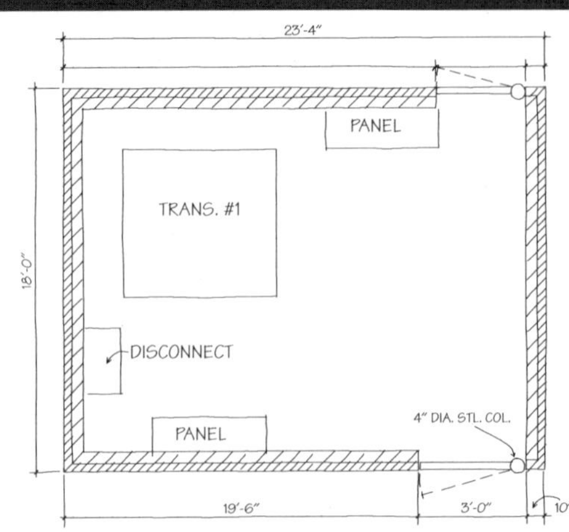

Figure 3-4. A sketch makes a good reference tool for a crew member who may forget a detail.

Penalties. Penalties incurred when a project runs over schedule hurts both the field leader and the contractor. Particularly on many large projects, if a contractor fails to meet deadlines in the contract, financial penalties will be assessed upon the completion of the job. The field leader who struggles with anticipating the direction of a project, who is behind schedule, and who is forced to catch up throughout the entire job will incur penalties.

When a field leader falls behind and poor timing ends up affecting other trades, they may not be so understanding. **See Figure 3-5.** Trades that are unable to proceed with their work, and can prove it was because the field leader did not finish a phase of work according to the established schedule, may back-charge the contractor for decreased production.

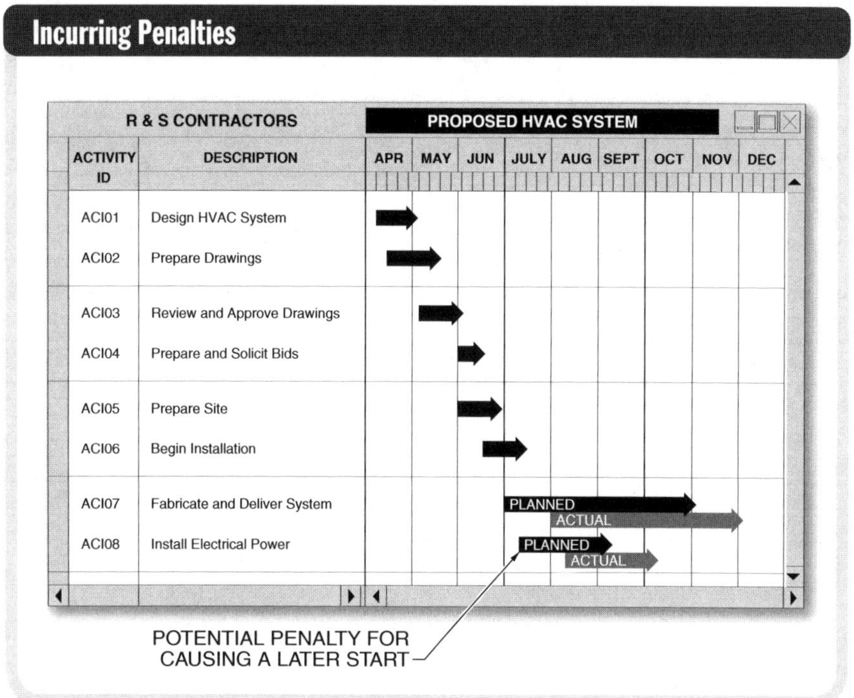

Figure 3-5. When a field leader falls behind and poor timing ends up affecting other trades, they may not be so understanding.

Loss of Credibility. Credibility is a lot harder to gain than it is to lose. Credibility is earned by following through on promises to the crew, to other trades, and to the customer. A field leader who makes promises but constantly comes up short will soon lack credibility. On the other hand, people tend to understand if on occasion something unexpected happens that leads to a missed deadline or commitment. If people know they can count on the field leader, they will generally be more willing to cut the person some slack. As long as a field leader continues to tell the truth and not just what everyone wants to hear and has established a track record for being credible, others are much more likely to be lenient. Credibility is built one step at a time, one job at a time, and it is an integral part of a field leader's career.

Loss of Respect. Respect is earned through action, not words. Field leaders ask their crews to be efficient and productive. If they do not see the field leader approaching the job with the same intent, their respect will ultimately disappear. The saying is old but true—one must lead by example. The better the example set for a crew, the more respect they will have for the field leader and the harder they will work.

Loss of Future Opportunities. A field leader who does not do a good job will probably not have a job for long. For example, if a lack of planning has led to a number of less-than-satisfactory projects, the contractor will not be happy with the results. If the company wants to keep the field leader, they may offer leadership training. If that effort fails, they will either demote or fire the field leader. The hard reality is that most field leaders do not receive training and are not demoted; instead, they are simply let go. Learning how to plan keeps a field leader's career in the best shape for long-term success.

Section 3.1 For Discussion

1. What is the purpose of making detailed lists?
2. How does a failure to plan directly affect a field leader?
3. How does a failure to plan directly affect a project and therefore the bottom line?

Section 3.1 Take Action

1. Create a list of the tasks required to complete a current installation or potential project.
2. Exchange task lists with a partner and identify potential problems that could occur during execution of the tasks listed.

Section 3.2

USING A PLANNING PROCESS

Just like each craftsperson working on a job site, each field leader will develop a personal approach to planning. Some field leaders inherently understand the importance of developing a comprehensive project plan, while others think thorough planning takes too much valuable time. Whether or not planning comes naturally, it is imperative for a field leader to understand that detailed project planning is not optional. Employing planning skills allows a field leader to create workable plans for any project, no matter the size or complexity of the installation. With experience, effective field leaders eventually develop an organized and repeatable planning system that creates predictable results.

Although every field leader puts a personal touch on every plan, each project plan involves the four phases of planning: detailing; tools and materials; manpower; and coordination. Identifying and carrying out each phase of a planning process ensures a functioning project plan, with no part being more important than another. **See Figure 3-6.** A project plan includes the sequencing of all the installations, each of which may require its own plan. For example, an electrical project plan may also encompass individual plans for utilities; feeder conduits; wall and ceiling rough-ins; lighting; feeder cables; branch conductors; and fire alarm systems. Regardless of the plan, if any one of the four phases is not addressed, the results will be subpar. Effective field leaders develop plans that make managing projects as efficient and productive as possible and yield successful projects that are completed on time and within budget.

A Four-Phase Planning Process

PHASE 1 — DETAILING

- Study project prints and documentation thoroughly.
- Identify questions that need to be clarified and complete RFIs.
- Make lists of both large and small tasks.
- Prioritize all tasks.
- Identify potential conflicts with other trades.

PHASE 2 — TOOLS AND MATERIALS

- Requisition tools and materials that must be on site before assembling a crew.
- Order all of the tools and materials that will be needed for the listed tasks that will be completed in the near future.
- Stock tools and materials where they will be easily accessible to the crew.

PHASE 3 — MANPOWER

- Use the list created during the detailing phase to assign hours, days, or weeks to each task.
- Use the project schedule to anticipate crew sizes and dates when they will be needed.
- Discuss manpower needs with the project manager prior to needing help.

PHASE 4 — COORDINATION

- Before coordinating with other trades, know what the tasks entail.
- Work together to find solutions.
- Be firm but fair when negotiating alternatives.
- Be as flexible as possible.

Figure 3-6. Identifying and carrying out each phase of a planning process ensures a functioning project plan, with no part being more important than another.

Phase 1—Detailing

Detailing is the critical first step of the planning process. This first phase of planning involves removing as much guesswork as possible from each task that will be assigned to a crew. To maximize production, the field leader needs to identify all the information, tools, and materials needed to schedule the execution of each project task.

DETAILING

- Study project prints and documentation thoroughly.
- Identify questions that need to be clarified and complete RFIs.
- Make lists of both large and small tasks.
- Prioritize all tasks.
- Identify potential conflicts with other trades.

A field leader's knowledge of the prints, specifications, and the project as a whole, as well as the crew's knowledge, has limits. It always helps to write things down. There is nothing worse than getting to a point in a project and realizing that more information is necessary to complete an installation. And it can be even more frustrating to realize that an issue noted in the past, yet never handled, is the root cause of the problem at hand.

Seek Answers. When detailing a plan, there are three important goals to accomplish. The first goal is to eliminate as many questions as possible before work begins, such as the dimensions, the materials to be used, the approach needed, and so on. It is important to examine every part of the prints to determine the best possible way to make each installation. During this part of the planning process, problems such as incomplete information, additional questions about the installation, and conflicts with other trades will likely arise.

For example, it may be necessary to request an interpretation from an architect or engineer to clarify the intent. A request for information (RFI) is used when directing questions to architects and engineers. **See Figure 3-7.** Unfortunately, it is not always possible to get immediate responses to issues or questions. Sometimes it can take weeks for them to consult with another architect or engineer who helped with the design, or with the customer, before responding to an RFI.

Document Answers. Once answers to all of the key questions have been received, it is vital to document the answers so that everyone working on the project can find them. Each set of prints that the crew will need to reference in the field needs to be kept updated with the latest information and revisions. Detailed documentation keeps everyone on the same page. If crew members are left in the dark, the installation will become riddled with errors, and valuable time and money will be wasted.

Work with Other Trades. Another part of detailing requires the field leader to work with other trades on a project. Working well with other trades helps eliminate the chance of having to lay something out numerous times. A field leader who helps other trades be efficient and productive earns credibility and respect from the field leaders of those trades. An effective field leader does not wait for the field leaders of other trades to ask scheduling questions, but is proactive and drives the job whenever possible. A field leader who waits for someone else to take the lead will end up working for someone else.

Requests for Information (RFIs)

Request For Information

Project: City Center Hospital **Sub-Contractor #** **RFI#** 35

Spec. Section: N/A

Drawing Reference: E101 One-Line Diagram and Feeder Schedule

Brief Description

Conductor sizes listed for feeder #F21 do not conform to the National Electrical Code.

Date Reply Request: 2/11/20XX

Critical to Schedule	**YES**	X	**NO**	
Cost Impact	**YES**	X	**NO**	

Submitted by: ABC Electrical **Date Submitted:** 2/4/20XX

Revision # **Date Submitted:**

Question(s)

The feeder schedule on sheet E101 calls for feeder F21 to consist of 4-250 MCM conductors and 1-#4 ground conductor run in 3" EMT. NEC requires that based on the 300 amp rating of the breaker, the phase conductors should be 300 MCM (article 310-16) with a #4 ground (article 250-122). Please advise. We anticipatethat the F21 conduit run will be approximately 150'. There will be a small increase cost for the increase in wire size. 4 x 150' moving up in size from 250 MCM to 300 MCM. The conduit size will stay the same.

Architect/Engineer/Owner Representative

Reviewed by: Architect Representative **Reviewed by:** Electrical Engineer

Signed: **Signed:**

Date Reviewed: 2/6/20XX **Date Reviewed:** 2/9/20XX

Reply:

Feeder F21 should consist of 4-300 MCM conductors and 1-#4 ground. Please proceed and forward a change order proposal for the increase in cost related to the change in wire size.

Figure 3-7. A request for information (RFI) is used when directing questions to architects and engineers.

Continuously Update the Plan. Each day that a field leader does not detail the plan, the crew gets a step closer to becoming aware of the lack of a solid plan that will lead to success for everyone. Detailing is best done as far ahead of time as possible, especially at the beginning of a project. This can really make a difference when the job is in full swing, and the field leader is being pulled in a million different directions. Some part of each day needs to be set aside to detail a specific part of the job. Slower times are the best times to detail, because detailing requires time to study and focus without being interrupted.

Ask for Help When Needed. Once a project is underway, it gets harder and harder to detail the plan. If at any point it becomes too much to handle, an effective field leader finds a way to get it done or asks for help. An effective field leader communicates with the project manager when help is needed. The project manager may be able to offer support in the form of more manpower or help prioritize what needs to be done.

Phase 2—Tools and Materials

> **TOOLS AND MATERIALS**
>
> - Requisition tools and materials that must be on site before assembling a crew.
> - Order all of the tools and materials that will be needed for the listed tasks that will be completed in the near future.
> - Stock tools and materials where they will be easily accessible to the crew.

Identifying the tools and materials that will be needed on site throughout the many stages of the project is the next step of the planning process. Field leaders make requisitions that are detailed enough to cover all the tools and materials needed for each stage of the project. It is absolutely crucial to have tools and materials on site before the crew needs them. The lack of even one item will bring the crew to a standstill.

With some projects, crew members are rarely within eyesight of the field leader. For this reason, the field leader may not be aware of the hours thrown away when the crew cannot find the tools or materials they need to complete their tasks. Unless a crew member speaks up, the field leader may never know. It is the field leader's responsibility to prevent wasted labor hours.

It is impossible to think of every single thing required for an entire project, but it is a reasonable expectation to have everything needed for the tasks planned over a three-week period. Following are a few simple tips for ordering tools and materials that save time and money.

Commonly Used Inexpensive Items. If 5000 bolts are needed throughout the length of a project, a field leader should not order bolts 200 at a time. The crew could easily use that many in a day and then stand around while the field leader spends valuable time running around trying to borrow bolts from another trade. There is no way to salvage wasted labor hours resulting from the crew not having the tools they need. **See Figure 3-8.** An effective field leader is mindful of purchases and tries to keep waste to a minimum. A field leader also recognizes that the cost of wasted labor can add up to a much higher number than the cost of a box of leftover bolts at the end of a project.

Quantities of Materials. An effective field leader knows it is best to order larger quantities of materials so that the crew will have what they need. If there is room to store large quantities of materials, doing so can save money. The fact that materials may be taken by other crews is not a good excuse for ordering smaller quantities. If there is concern that other trades may help themselves to the materials, the field leader needs to provide gang boxes and give keys to the crew members. An effective field leader also makes sure the crew knows how important it is to keep the gang boxes locked at all times.

Specialty Tools and Materials. There will always be certain tools and materials that are not readily available and in stock at a supply house. As soon as the field leader identifies those items, a plan and a schedule is developed for having those items on site. There is nothing worse than having a crew ready and willing, only to discover that the specialty tool needed will be unavailable for a week. If the crew does not have the tools they need, they may be unable to complete the work scheduled. When this happens, every trade on the project will find it tough to get their work done on time.

Missing Tools

Figure 3-8. There is no way to salvage wasted labor hours resulting from the crew not having the tools they need.

Leading in the Field: Figuring Burn Rate

As a field leader, in order to truly understand the consequences of wasted labor, it is important to know each crew member's actual "burn rate" per hour. First, take the amount each individual makes per hour and then add all the benefits received beyond that hourly wage. Next, divide the individual's hourly total income by 60 minutes. This number is the dollar value per minute that can be placed in the "red column" every time a worker wastes time searching for a tool or material that should be readily available. Tools like extension cords, socket sets, and ladders are so inexpensive in relation to wasted labor that a crew member should never have to go far to find them. It is always smart to have plenty of tools like these at any job site. The extra expense is worth it.

On-Site Deliveries. Heavy competition among supply houses has changed the way business is done, particularly in large market areas. When suppliers are willing to do whatever it takes to get an order, field leaders can become complacent by assuming they can get what they want immediately.

When a project is in a location that does not have easy access to supply services, a field leader must plan ahead. If a field leader is constantly ordering small amounts of materials because they are about to run out, then the crew will run out of what they need anyway at some point. This is why habitually ordering only for the next day or two will keep the project perpetually inefficient. Plus, every time a delivery comes to the site, someone has to stop work to sign for it and then unload the truck. Then the materials have to be taken to the on-site storage area. If deliveries are made only once a week and include materials for the entire week, they are much more efficient.

Every job site is different in terms of how deliveries are handled. Some sites assign delivery times to avoid bottlenecks when multiple trucks show up at the same time. Sometimes, it can be logistically challenging to get materials and tools to where they are needed. Some deliveries require forklifts or cranes to be unloaded and need specialized attention and planning. No one likes to have a delivery show up with no way of getting it off the truck.

When deliveries flood in without a schedule, crew members are much more likely to take what they need and push the rest aside. Many of these materials tend to get lost in all the activity and are never seen again. It is the field leader's responsibility to put a system in place for getting materials where they need to be.

Phase 3—Manpower

Once all the tools and materials have been handled, it is time to proceed to the next phase of the planning process and determine how large a crew is needed for the project. The field leader also needs to determine when it will be necessary to increase or decrease the number of crew members. A detailed plan provides an excellent guide for building an accurate manpower chart.

> **MANPOWER**
>
> - Use the list created during the detailing phase to assign hours, days, or weeks to each task.
> - Use the project schedule to anticipate crew sizes and dates when they will be needed.
> - Discuss manpower needs with the project manager prior to needing help.

Charting Manpower. The field leader must discuss the manpower chart with the project manager and then build a schedule to ensure that help will be available when needed. **See Figure 3-9.** The project manager and the field leader need to agree on when to increase or decrease the number of crew members. Keep in mind that the project manager probably built a manpower chart during the design or estimating process. A field leader who makes the mistake of adding people whenever they are perceived to be needed runs the risk of putting the project's success in jeopardy. Just like the rest of the project plan, a manpower chart is a necessary reference tool that is also a work in progress.

Adjusting Manpower. Wasted labor hours cost a project more money, and at a faster rate, than anything else. That is why it is up to the field leader to make sure the number of crew members is the right number for the job. As soon as there are too many crew members or not enough crew members for the remaining amount of work, then it is time to adjust manpower.

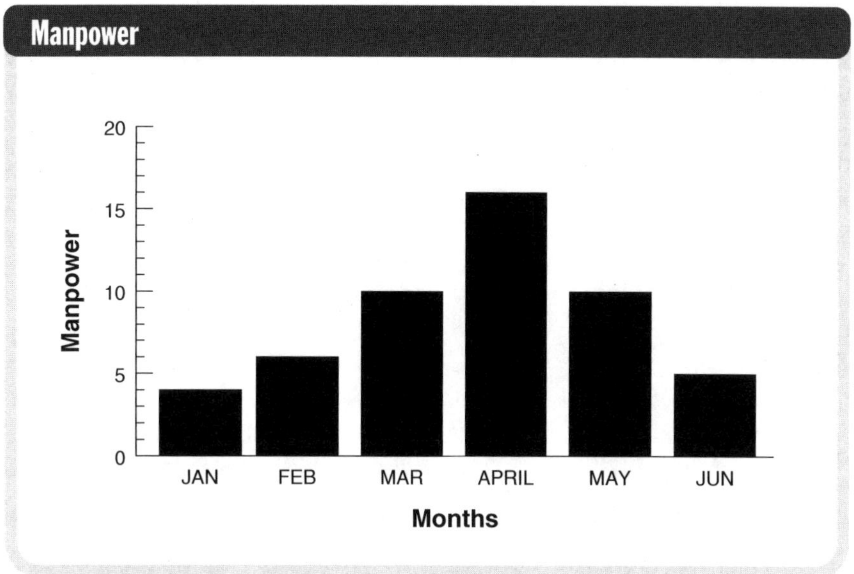

Figure 3-9. The field leader must discuss the manpower chart with the project manager and then build a schedule to ensure that help will be available when needed.

Leading in the Field: Too Many Hands

A field leader is waiting impatiently for another trade to finish their part of an installation and does not have enough work to keep the crew busy and productive every day. Because the field leader is fearful that the crew will quit, busy work is assigned instead of finding work for them on another project. What is the result? Valuable labor hours are wasted, which endangers the project's success.

Too Many Crew Members. The trap of keeping too many crew members as things begin to slow down is a common one. It is every field leader's fear that as soon as part of the crew is sent home, there will suddenly be too many tasks to complete and things will pile up. When cuts need to be made, an effective field leader knows that temporary help can always be accessed if absolutely necessary. And the labor hours saved may be needed later in the project. If those hours are not needed, they will be added to the project's profits at the end.

Too many people on a crew can also damage efficiency. It is easy to lose sight of the bigger picture when this happens and to imagine that it will be helpful to have an oversized crew spend time finishing all of the small incidental tasks that have been sitting on the list of things to complete, like finish work and labeling. But this stopgap measure rarely works. If a field leader continues to use a large crew to complete smaller tasks, then the work will run out at some point. As much as these things need to be done, they generally can be accomplished at the end of the job with a smaller crew. This does not mean that it is good to wait until the last minute. Rather, an effective field leader trims the crew while waiting for the other trades to finish and has a small crew work on the list of smaller projects in the meantime.

Not Enough Crew Members. There is also cause for concern when the number of crew members is cut back too soon. It is all about finding and maintaining balance. Early cuts may seem like a good idea because they lessen the number of labor hours spent. The downside of this maneuver is that it often results in missed schedules and a scramble to complete tasks. Ultimately, the goal is to meet the schedule so the other trades who follow will not have to scramble to make theirs.

Leading in the Field: Cutting Back Too Early

A field leader has cut back the crew too early, and the project is behind schedule. The field leader cracks the whip on the small crew that remains, but they can only do what they can do. The field leader comes across as having unreasonable expectations, but continues to push the crew even harder. The crew hits a snag on one of the tasks, and the field leader cannot move forward with the other tasks on the schedule. Both the field leader and the crew become even more stressed.

Phase 4—Coordination

Once a plan is in place, it is important to recognize that a field leader's plan does not stand on its own. Working with field leaders of the other trades to be sure the plan works for them is the next phase in the planning process. The field leaders of other trades will also be planning what they need for their portion of the project. Field leaders who work together to coordinate plans will identify conflicts, which makes it easier to remedy them prior to the onset of installations.

COORDINATION

- Before coordinating with other trades, know what the tasks entail.
- Work together to find solutions.
- Be firm but fair when negotiating alternatives.
- Be as flexible as possible.

Communication is Key. Communication is key to successful coordination efforts. When a field leader is at a job site and the project manager and support staff are at other locations, it is impossible for them to know what they can do to help each other unless the field leader communicates. Good communication means giving everyone as much notice as possible so they will not be blindsided by requests. Most project managers and support staff are responsible for many different projects that are running simultaneously. The field leader who always calls at the last minute with an emergency not only loses credibility with the project manager but is not likely to get help when a situation warrants it.

Partnerships with Other Trades. Working with other trades means forging partnerships. On projects where the trades do not work well together, making a profit becomes difficult. It often becomes a battle to see who can get the work done first, which causes tension and resentment. A trade that believes their work should take priority creates dissension and discomfort on the job site. Instead, finding a common goal improves everyone's chances for a profitable outcome.

An effective field leader makes sure to give input before the project schedule is formed. **See Figure 3-10.** If input is not given, the field leader will constantly be reacting to the actions of the other trades. Essentially, the field leader will have given the other trades the right to put their interests first. The field leader of each trade is only responsible for the success of that trade's portion of the project. Effective field leaders make sure the schedule is coordinated with input from each trade.

Proposed Schedules

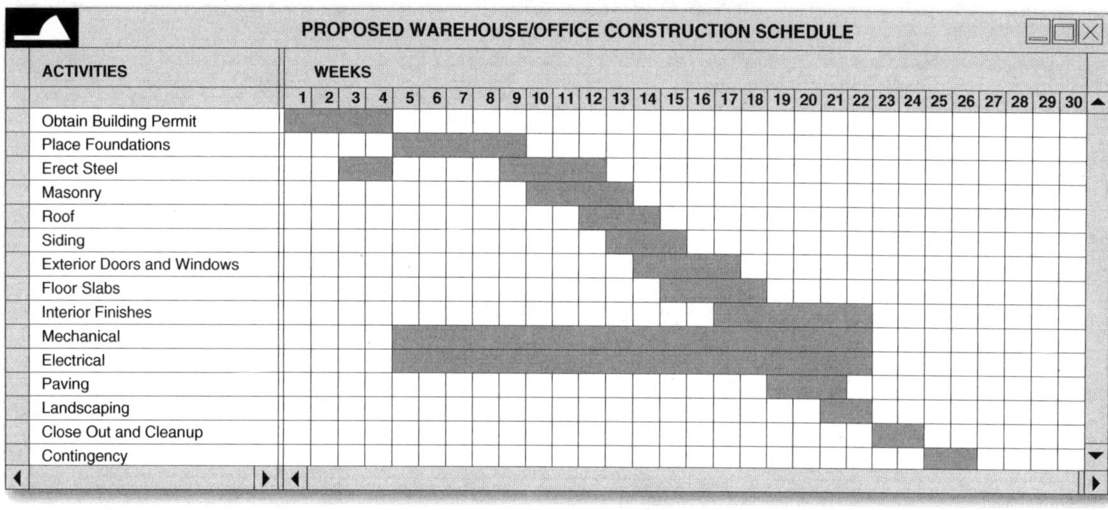

Figure 3-10. An effective field leader makes sure to give input before the project schedule is formed.

Section 3.2 For Discussion

1. Explain how each phase of the four-phase planning process is used.
2. How is an RFI used during the planning process?
3. Explain why it is best to order larger quantities of materials for a project.
4. Discuss the drawbacks of having too many crew members for a project.
5. Why is communication important when coordinating with other trades?

Section 3.2 Take Action

1. Complete an RFI.
2. Role-play how different trades need to communicate during detailing and coordinating.
3. Use the four-phase planning process to write out a project plan.

Section 3.3 PLANNING CHALLENGES

Inherent to any job are challenges that need to be overcome in order to be successful. It can be unsettling to learn that many of the challenges a construction field leader faces were not covered in apprenticeship training nor pointed out by foremen in the past.

It is not unusual for a field leader to be overwhelmed by the number of tasks and responsibilities that come with the job. However, remaining overwhelmed will compromise a field leader's effectiveness. When this happens, the ability to stay focused may be lost. An effective field leader faces challenges head on and learns how to overcome them. The following guidelines offer suggestions for handling planning challenges that may arise on a project.

One Thing at a Time

Being a field leader requires the ability to respond to multiple tasks, seemingly at the same time. However, regardless of the number of tasks, the primary focus of a field leader should always be on the project plan. By focusing on the plan, the field leader is less likely to make serious missteps that can lead to rework and omissions during installation. **See Figure 3-11.**

Focus on the Plan

Figure 3-11. By focusing on the plan, the field leader is less likely to make serious missteps that can lead to rework and omissions during installation.

It is important to remember that it is always best to tackle one thing at a time. Multitasking minimizes the total effort that can be put into each individual task, leading to a less-than-desirable result. Multitasking may appear impressive—handling a phone call, writing an email, and gesturing to someone all at once. But the reality is that the person on the other end of the phone likely feels slighted, the email probably contains errors, and the gesture likely confused the person on the receiving end.

But finding the time to focus on the plan is another matter, especially when people are demanding the field leader's time in order to get their jobs done. That is why the field leader's approach to this assignment needs to be the same as for the crew members. A field leader is least likely to be disturbed or distracted very early or very late in the day. These are the best times to focus on planning. At other times, the field leader will be busy working out issues with the crew.

Make Notes and Lists

Keeping track of everything that happens on a job site without forgetting any of it—or even half of it—is an unlikely prospect. Yes, everyone is different, but no one is without limitations. A field leader who does not write anything down may forget to order materials the crew needs for the next day or may forget about a commitment made to a customer. Either way, forgetting things brings unwanted results.

Notes and lists provide an excellent way to focus attention and energy where it is needed most. What materials or tools are required? What must be done? Who is involved? All these things should end up on notes and lists to be handled during the time set aside at the beginning or end of the day. A list will continue to grow even as tasks are accomplished. But if a field leader keeps a master list that is always revised, based on new and completed tasks from the crew and other trades, the field leader will always feel that things are under control.

The goal is to never be without a way to record new issues that crop up while walking the job site. A small pocket notebook is a great solution, and the items jotted down can easily be transferred to a master list. A master list can be kept in a printed notebook or as digital files. **See Figure 3-12.** In fact, just the act of transferring the information at the end of the day can help a field leader reorganize job tasks for future planning purposes.

Figure 3-12. A master list can be kept in a printed notebook or as digital files.

Another option is to make notes on a digital recorder or smart phone. It is easy to hit the record button and talk for a few seconds while walking away from a conversation and the thoughts are fresh. Not only is it easier to finish thoughts without being interrupted, but they do not have to be written down. And since digital recorders and smart phones date and time-stamp notes, those do not have to be written down either. Another plus for recording devices is that they can be taken into coordination meetings, which allows a field leader to focus on the meeting rather than on taking notes.

A field leader who does not make notes and lists, and is constantly in a reactionary mode because of it, is doing the crew, the company, and the customer a disservice. Each time a field leader is held accountable but does not come through, credibility is lost.

Prioritizing

Set Priorities

When directing a crew throughout a project, a field leader always attempts to complete the highest priority tasks first to keep ahead of schedule and avoid getting backed into a corner. No matter how long or short the list of tasks, an effective field leader takes the time to number the tasks in order of priority. An effective field leader also updates the list frequently. Also, if the crew is completing a task that could be done at a later time, it may inadvertently eliminate part of the contingency plan put in place for when the project slows down or hits a snag. But this does not mean avoiding tasks that need starting or leaving tasks unfinished for the sake of saving them for later.

No matter how long or short the list of tasks, an effective field leader takes the time to number the tasks in order of priority.

It is better to have a small crew working on priority projects than to have a large crew completing tasks that a small crew can handle near the end of a project. Effectively managing what needs to be done with a small crew means banking labor hours that, if needed, can be used at a later point. An effective field leader always makes decisions about manpower based on what is best for the project in the long run.

Minimize Interruptions

Interruptions are as unavoidable as hard hats in the construction industry. That is why it is important for a field leader to use methods to help remember exactly what was being done when a distraction occurred. It is common for a field leader to receive calls over the radio or phone multiple times a day when the crew needs input on a task. It may require a walk to where the crew is working, or it may be as simple as clarifying something verbally. Either way, it breaks the field leader's concentration. It is also commonplace to have the field leader from another trade ask to walk the job to discuss upcoming coordination efforts.

When an interruption takes place, a field leader must make a conscious effort to get back to the previous task. For example, the field leader may be studying prints when a delivery driver walks through the door. The driver doesn't know where to go and is blocking traffic with the delivery truck. In this case, a sticky note can be used to remind the field leader of what was going on before the interruption, marking both the place the field leader stopped on a given document and the thoughts associated with it. A field leader can never make too many notes. And if the field leader waits until the next day to make notes, there is a good chance they will be forgotten.

Some interruptions are the result of personal choices. For example, using electronic devices for personal use and gossiping lower productivity, lead to schedule disruptions, and can cause conflicts. The following tips can help minimize these types of interruptions.

Personal Calls. With a cell phone, everyone has the ability to call anyone at any time or to take calls from anyone at any time. The problem begins when personal phone calls encroach on time at work. Everyone has personal issues outside of the job, and occasionally they may require attention with a phone call in the middle of the day. But, most calls are not urgent. Every personal call leads to a loss of productivity and, in turn, an unplanned project cost. **See Figure 3-13.**

Personal Calls

Figure 3-13. Every personal call leads to a loss of productivity and, in turn, an unplanned project cost.

Some field leaders argue that because they frequently put in more time than they are getting paid for, they have earned the right to use work time for personal matters. But here is the reality. Even if a field leader puts in more time, that time needs to be maximized for the express purpose of doing extra planning.

Leading in the Field: On the Job

A contractor comes to your house to fix a plumbing problem. This contractor gets paid by the hour. Instead of being in the bathroom, the contractor is standing by the truck talking on the phone. Half an hour later that contractor is still talking on the phone. Is this on a personal call? Is the conversation related to the job? The contractor gets paid by the hour. Should the contractor be paid for the half hour spent on the phone?

There is another problem with taking and making a lot of personal calls during work hours. Everyone has a certain sense of entitlement when putting in extra time that is not reflected in a paycheck. But it is not a question of entitlement. The crew will not see the extra time a field leader puts in because they come to work later and probably go home earlier. A field leader should not be seen making or taking personal calls, this gives the crew members license to do the same.

Surfing the Web. Time can slip away faster online than in almost any other activity. Having online access to research information that is job related and the ability to send and receive email is an advantage on any project. Browsing personal interests online after looking up something job related is a tempting distraction. It is easy to get sidetracked, and before long, a chunk of time will have disappeared in the blink of an eye. An effective field leader either closes the browser or turns off the device as soon as the answer to the job-related question is found and gets back to the planning tasks at hand.

An effective field leader knows how to manage the time spent online during work hours. Checking emails or gathering online information at the beginning or end of each day may be a solution. The field leader also avoids social networking sites while on the job because there is no such thing as a short visit on social media.

Gossiping. It is inappropriate for a field leader to gossip with the crew or crew members of other trades. Gossip only spreads rumors, hurts feelings, and leads to speculation that results in unproductive downtime. When production is slowed by one or more workers, the time has to be made up somewhere else by someone else—maybe even by the field leader.

Section 3.3 For Discussion

1. What are some common planning challenges?
2. Why is multitasking a poor approach to completing tasks?
3. How does setting priorities help a field leader stay focused?
4. Name common interruptions that often occur on the job site.
5. Why should a field leader limit making personal phone calls during work hours?

Section 3.3 Take Action

1. Role-play how an overwhelmed field leader can do one thing at a time to stay focused.
2. Make a list of tasks for a current or potential project and prioritize those tasks.
3. Role-play how a field leader can effectively minimize interruptions.

Section 3.4

THE IMPORTANCE OF ORGANIZATION

What is the difference between planning and organizing? Planning is the process used to ensure that the necessary tools, materials, information, and manpower are on site when needed. Organizing is the process of making sure that the appropriate tools, materials, and information are accessible to maximize production. It is the responsibility of the field leader to make sure that the crew works as productively and efficiently as possible. Therefore, it is also the responsibility of the field leader to make sure the project stays organized.

To keep a project organized, the field leader must have systems in place that promote an organized job site and must communicate organizational expectations clearly and frequently. A well-run project takes a concentrated effort by the field leader to put systems in place that enable the crew to start out organized and stay organized. Two things are necessary:

- A field leader must keep organization in the forefront as a key to the success of the entire project.
- A field leader must use consistent communication skills to empower the crew to do what is necessary to maintain an organized job site to maximize production.

Leading in the Field: Totally Disorganized

The job site is in disarray. Tools and materials often have to be tracked down before they can be used by crew members. Each day, the field leader says, "I know I ordered that tool. It has to be around here somewhere." The crew members are expected to track down missing tools. Each time someone needs a tool, the same thing happens—a crew member is sent around the job site to locate it. Finding a missing tool takes anywhere from a few minutes to more than an hour, and the lost time during production directly affects the bottom line.

Some field leaders think that being organized is someone else's job; others see it as unimportant. But this is simply not true. Bad habits, such as being disorganized, produce substandard results. Being organized requires focused and consistent attention. It is easy to be consumed by other responsibilities that require immediate attention. An effective field leader does not make the mistake of dismissing organizational issues. Organization is a key element of a field leader's long-term success.

Organized Job Sites

Job sites are often messy and disorganized, which can make them unsafe. Most construction workers believe it is just the way things are because that is what they have experienced throughout their careers. When crew members say, "How can we keep it clean when we are constantly surrounded by so much debris?" the field leader's answer should be, "It's not about keeping the site spotless. It is about keeping the tools, materials, and safety equipment in order and away from the work area, and removing any debris as soon as possible."

Every single decision a field leader makes has the potential to increase or decrease production. Just think about a crew member who is wandering the job site looking for a tool. If the crew member is lucky, the tool will be found quickly. But that is often not the case. Generally, a crew member will not volunteer how long it took to find the tool, and it will probably not be possible to determine how much time was spent looking for it. In this case, the field leader has inadvertently provided the worker with a free pass to roam the site searching for something that may not be found.

Furthermore, as the crew member looks for the missing tool and asks other workers if they have seen it, conversations about the upcoming weekend or about last night's game may arise. Other crew members may get into a heated discussion about the fact that they can never find the tools they need to get their jobs done.

At this point, a field leader may see a couple of crew members who appear to be doing nothing but talking and ask, "What's going on? I thought you were working on another floor." To which the crew member might tersely respond, "Don't you remember? You sent me to look for that missing tool. I haven't found it yet." As a field leader, it is important to remember that disorganization can lead to wasted labor costs and can directly affect crew morale, which in turn can decrease production.

The bottom line is that organization is a leadership skill. Every project presents yet another opportunity for a field leader to create a standard of organization that will ultimately maximize project production and success.

Locating Staging Areas. The day a field leader starts looking through the prints is the day sites for the staging areas for tools and materials should be considered. For greatest efficiency, the field leader must look for large, open staging areas that are located close to where the crew will be working. **See Figure 3-14.** Having staging areas as centrally located as possible cuts down the time it will take the crew to gather what is needed for a task. Consider the following when identifying staging areas:

- Does the staging area have easy access for deliveries?
- Are there any height or width restrictions that will create problems along the path from the delivery area to the staging area?
- Is the area located so that items will have to be moved multiple times so other trades can have access to complete their tasks during the project?

Once a staging area is selected, the field leader communicates that to the other trades and the general contractor. This gives others a chance to speak up if there is a reason why the selected area might cause a conflict. An effective field leader drives the conversation so as to not get stuck with staging areas that no one else wants and that leave the crew at a disadvantage.

Once a staging area is set, the field leader must move the gang boxes and material racks into place as soon as possible. A wide open space is an invitation for another trade to move in. And having a crew move to another location after they have claimed a staging area only creates the potential for friction.

Staging Areas

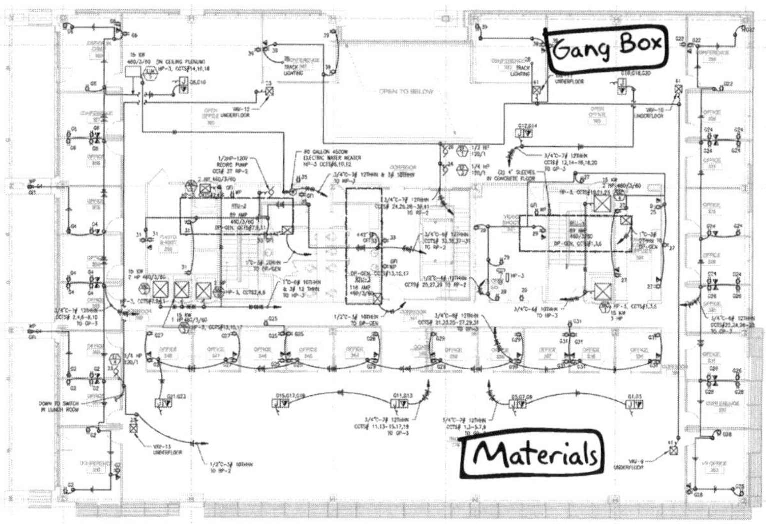

Figure 3-14. For greatest efficiency, the field leader must look for large, open staging areas that are located close to where the crew will be working.

Staying Organized. Next, the crew has to be convinced of the importance of keeping things organized. If a field leader can win a crew over to this way of thinking, the project will stand a much better chance of success. It is easier said than done, but do not give up. Stay firm in the position that a messy and disorganized site will not be tolerated. It may take many conversations and regular reminders, but do not back off on the commitment to remain organized.

Crew members bring with them as many good and bad habits and attitudes as there are individuals. They have picked up work-related routines learned from all their previous projects. It is not that they are unwilling to do what is being asked of them, but they may not be accustomed to a new way of doing it. A field leader must encourage them to believe that good habits can be accomplished through consistent communication and positive reinforcement.

One way of encouraging the crew and getting them to trust a new routine is to share a short, relatable story or experience that hopefully resonates with them. In this case, a field leader might say, "You may think I'm obsessed with organization because I comment about it at every crew meeting. My reasons for doing that come from when I worked as an apprentice and journeyman. It drove me crazy when I was asked to install something, and the tools or materials were nowhere to be found. It also stressed me out trying to find what was needed to complete the task. Because of those experiences, it is my goal to avoid doing that to anyone on my crew. I appreciate the contribution each of you make to the project."

Safe Job Sites

An organized job site creates a safe and positive work environment. Organizing safety equipment in a readily accessible location is a key step to running a safe job. When crew members know where to find the safety equipment and that it will be in good working order, they are more likely to use it. **See Figure 3-15.** It is that simple. On the other hand, if the crew believes that they are going to have to search high and low through the entire job site to find a piece of safety gear when they are starting a task, there is an increased risk that they may skip using the appropriate safety measures altogether.

Safe Job Sites

Figure 3-15. When crew members know where to find safety equipment and that it will be in good working order, they are more likely to use it.

Section 3.4 **For Discussion**

1. What two actions must a field leader take to keep a project organized?
2. Share a personal experience about not being able to readily access the proper tool for a task.
3. How frequently do crew members need to be reminded to keep things organized?

Section 3.4 **Take Action**

1. Role-play two exchanges on the importance of organization as viewed by a field leader and the crew members.
2. Use a job site print to designate two possible staging areas.
3. Organize a list of tools and materials by indicating where each can be safely accessed for the most efficient execution of tasks.

Planning and Organizing

Case Study

The first 15 floors of a 20-story hotel are mostly complete and are being painted and trimmed out. The top five floors are in various stages of framing and rough in. The field leader is in charge of an eight-person crew responsible for the electrical rough in of the top five floors.

Tuesday

1:00 pm The field leader walks the job and checks in with everyone on his crew. Everyone appears to have what they need, so the field leader decides to go back to the trailer to do some more detailing on the prints. On his way back to the trailer, the field leader gets caught up in a conversation with other trade field leaders and ends up spending most of the afternoon visiting.

Wednesday

6:30 am The field leader thinks about how the day should go as he commutes to the job site. Based on what he saw the day before and what phase the project is in, he concludes that the job should continue down a smooth path. He believes the hectic days of getting the job on track are finally behind him, and now it is just a matter of going through the motions to complete the project.

7:02 am The field leader has a quick discussion with the crew about safety relating to the tasks planned for the day. During the meeting, the field leader receives a call from a crew member saying he is sick and will not be coming in. The field leader quickly rearranges a few crew members to cover for the absence. As the crew heads out to the job site, the project manager arrives unannounced. He wants to discuss project timelines, billing progress, and project budget, so he and the field leader head to the trailer.

7:14 am The field leader receives a call from a crew member telling him that the manlift has broken down and the seven-man crew is still on the ground waiting for a ride to the upper floors. The field leader tells his crew to take the stairs. The project manager overhears the conversation and comments, "There go two wasted labor hours we'll never get back!" By the time the crew gets to work, it will be 7:30 am and 3.5 labor hours will have been lost.

7:27 am The field leader gets a call from the general contractor who believes the manlift problem is due to a power issue and requests his attention because it is holding up the whole project. The field leader is still meeting with the project manager, so he calls his most trusted journeyman to handle it. But in doing so, he pulls the journeyman away from an important task.

7:45 am The field leader gets a call from his most trusted journeyman who is troubleshooting the manlift. The journeyman asks, "Have you seen the digital voltage meter lately? I've checked all the gang boxes and I can't find it." The field leader then realizes that the meter is sitting on his desk. So he calls the material handler and tells him to come and get it and deliver it to the journeyman.

Case Study ...continued...

Wednesday

8:02 am

The field leader receives a call from a crew member who has run out of junction boxes on the 18th floor and wants to know when more are being brought up. The field leader remembers that, the day before, he meant to tell the material handler to stock the 18th floor by the end of the day. But the field leader did not make a note of the task, and it slipped his mind. The field leader calls the material handler and tells him to get the materials up there right away.

8:15 am

The material handler walks into the trailer where the field leader and project manager are meeting. He asks, "Do you know what happened to all those junction boxes we had delivered last week? I can't find them anywhere." The field leader begins calling the field leaders of the other trades to ask if they know. It begins to dawn on him that they may have all been used, and none are scheduled to be delivered. The field leader then makes an urgent call to the nearest vendor to see if they can deliver some in the next hour.

8:41 am

As the field leader discusses target completion dates with the project manager, he is reminded that there is a change order that was written four months prior and that the company still has not been paid. The project manager wants to make a copy of the original signed change order to take back to the office. The field leader stares at a file cabinet that he knows is extremely disorganized. He cringes because it is not going to be easy sifting through all the change orders, RFIs, and daily job journals that have been randomly stuffed in the file over the past nine months. He asks the project manager if he can get back to him in a couple of days.

9:00 am

After the project manager leaves the site, the field leader makes his way to the 16th floor to check on the crew's progress. He finds that one of his crew members has installed something incorrectly.

Field Leader:	"What's going on? This is in the wrong location!"
Crew Member:	"That's where you told me to put it! When we talked over the radio yesterday, you said to mount it next to the south side of the west window."
Field Leader:	"No, I said the south side of the east window!"
Crew Member:	"That's not the way I remember it! I know I've mentioned it before, but if a set of prints were available up here, I could've confirmed the location."
Field Leader:	"I need you to drop what you're doing and get that moved and rewired. The inspector is coming at 11 am, and I promised the general contractor that we would get inspected today so that the insulators and drywallers could start in here tomorrow."

Case Study ...continued...

9:11 am The field leader gets a call from the general contractor asking him to come up to the 17th floor. Once there, the field leader realizes that he failed to inform the crew that there was a change to the layout used on previous floors. This had been agreed upon in a coordination meeting with the HVAC contractor two weeks before. Two crew members are working in the area, so the field leader tells them to stop what they are working on and to remove the recessed can lights so the HVAC contractor's team can continue their work.

10:00 am The field leader finally has a moment to stop and think. "What in the heck happened today?! I had everything running smoothly, and in only a few hours, total chaos erupted. How am I supposed to plan for all of these unforeseen things?! I sure hope the rest of the day goes better, or I'm going to lose my mind!"

10:01 am The field leader sits down to gather his thoughts and grab a cup of coffee. The phone rings, and it is the inspector. He is an hour early and wants to walk the job.

TAKE A CLOSER LOOK

1. Which events occurred as a result of poor planning?
2. Which events occurred as a result of disorganization?
3. If any of the events were unavoidable, what could the field leader have done differently?

Case Study ...continued...

How Can an Effective Field Leader Respond to Unplanned Events?

If a crew member calls in sick:

- Pull manpower (when practical) from tasks that are of the lowest priority.
- Make sound decisions when redirecting manpower.

If a project manager arrives unannounced:

- Be professional and remain positive.
- Focus on completed tasks and ask for support if needed for uncompleted tasks.

If the crew waits for the manlift to be fixed before going to work:

- Explain to the crew that they should take the lift when it is available, but if there is a backup for any reason, they should take the stairs.
- Take time to explain to crew members how labor costs affect the project overall.

If manpower is needed to fix the manlift:

- Prioritize tasks in order to make sound decisions about where to pull resources.
- Allocate appropriate manpower to fix the manlift as soon as possible.

If a special tool (digital voltage meter) is misplaced:

- Order additional specialty tools if needed in more than one location (floor) at the same time.
- Communicate with the crew regarding the importance of putting specialty tools away so that anyone can find what they need.
- Set a good example for the crew by maintaining an organized job trailer.

If a crew member runs out of materials:

- Place a notepad near the gang box so crew members can jot down requests.
- Order materials listed (on the notepad) as soon as possible.

If there is a shortage of materials:

- Communicate appreciation to crew members for the help they provide by reporting when materials are running low.
- Strive to keep materials organized and staged in the same place so they are much easier to quantify during a walkthrough.
- Make it a point to ask, as the crew is locking up, if anyone is getting low on materials or tools, or if they can think of anything else they may need in the immediate future.

If an incorrect installation is made:

- Be professional and respond calmly.
- Provide face-to-face instructions for correcting the mistake.
- Encourage crew members to ask for clarification to ensure accurate installations.

Case Study ...continued

PERSONAL PERSPECTIVE

1. Describe an unexpected situation that arose on a job site and how a solid project plan could have been used to resolve that situation.

2. Describe how a lack of organization impeded work on a job site and how organizational skills could have been used to resolve the situation.

3. Role-play an unplanned event that occurs on a job site and how it can be avoided by having a solid project plan and an organized job site.

chapter

4

PROJECT COORDINATION

It is a field leader's responsibility to coordinate all the activities on a project to make sure that tasks are completed safely, according to code, on time, and within budget. A field leader also has the responsibility of directing and training the crew so that the quality of their work will meet the expectations of the customer and must resolve any problems that arise and document them appropriately.

A field leader addresses production challenges on a daily basis and continually prioritizes tasks to ensure that the crew operates at maximum efficiency. An effective field leader constantly moves high-priority tasks to the top of the daily job list while monitoring progress on longer-term tasks that are lower on the list. Project coordination is quite stressful, and an effective field leader takes steps to minimize stress and establish a work-life balance.

Section 4.1: Safety First
- Explain why field leaders must always "walk and talk" safety.
- Explain why risky behaviors cannot be tolerated on a job site.

Section 4.2: Documentation
- Explain how keeping accurate records aids the compilation of a final project report.
- Describe how change orders, daily job logs, RFIs, purchase orders, packing slips, as-builts, job progress, and coordination meeting action items should be documented.
- Establish a filing system for project documentation.

Section 4.3: Production Challenges
- Identify 10 common production challenges that a field leader may face on a project.
- Describe how an effective field leader addresses the 10 common production challenges.

Section 4.4: The Stress Factor
- Explain how identifying the sources of stress helps a field leader lower stress.
- Explain how a field leader's reasonable expectations are key to self-preservation.
- Describe the role that work-life balance plays in a field leader's effectiveness.

Learner Resources
ATPeResources.com/QuickLinks
Access Code: **335421**

Section 4.1

SAFETY FIRST

Construction work is, by nature, a physically demanding and dangerous vocation. It takes everyone, from the top on down, to do their part to make sure that injuries and accidents are avoided at all costs. It is the responsibility of the field leader to make sure that the crew has whatever they need to remain safe and injury free on the job site. The crew relies on their field leader to provide safety gear and equipment, the necessary tools and materials, and the information needed to complete each task.

By communicating safety expectations and providing the crew with the safety equipment they need, an effective field leader makes it clear that safety on the job site is important.

An effective field leader makes it known that anything less than the crew doing everything in their power to be safe on the job will not be tolerated. By communicating safety expectations and providing the crew with the safety equipment they need, an effective field leader makes it clear that safety on the job site is important. The attention a field leader pays to having the crew complete every task safely is just as important as a detailed project plan. It may be as simple as making sure the crew always wears hard hats or requiring the crew to be in safety harnesses and tied off when working at the leading edge of a building. No matter how big or small, how obviously dangerous or seemingly secure, good safety practices cannot be assumed or overlooked.

Safety Leadership

A field leader is responsible for making safety expectations very clear—safety first—and for making sure crew members understand that no one will be penalized for doing things safely. If an individual does not feel comfortable asking for safety equipment, it is a problem. A crew member who takes a laid-back approach, thinking there will be no consequences if the field leader sees the unsafe practices, is an accident waiting to happen.

How a crew perceives a field leader's commitment to safety determines how they will react when met with an unsafe situation. Some field leaders give pretty convincing speeches at safety meetings, but then they turn around and either completely ignore safety protocols or ask their crews to perform tasks without taking the time to adhere to safety protocols. The crew will determine, in their own minds, if the field leader operates a safe job site. If a field leader consistently falls short of the safety image given, the respect and credibility that may once have existed will be lost. Effective field leaders both "walk and talk" safety.

An effective field leader sets up organized methods for storing and maintaining safety equipment on each job. **See Figure 4-1.** However, if a field leader allows the crew to abuse safety equipment or store it incorrectly, the chance increases that the equipment could become faulty. A clear and consistent message from the field leader explaining what is expected in terms of maintaining safety equipment provides a basis on which the crew can be held accountable.

Safety Equipment

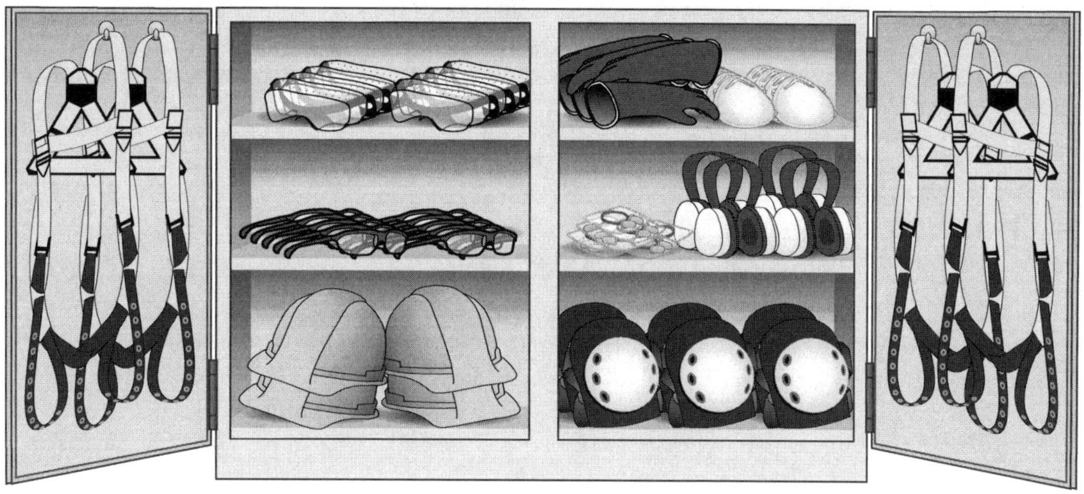

Figure 4-1. An effective field leader sets up organized methods for storing and maintaining safety equipment on each job.

It is an unfortunate reality, but some field leaders believe there is already too much to do in terms of keeping their crews productive, so they regard safety as just something else to slow them down. To argue this point is senseless. If a field leader's crew does not work safely, there is a good chance the field leader will be without a job or in a courtroom defending a negligence suit. It is important that a field leader recognize that there are far-reaching consequences for choosing to look the other way when it comes to safety. Ignoring responsibility when it comes to safety is a huge personal risk.

Communicating Safety

A lack of communication between a field leader and crew can directly affect everyone's safety. Part of a field leader's job is to create a safe work environment and to hold crew members accountable when they take unnecessary risks. A field leader cannot set the precedent of looking the other way for the sake of production. If a field leader ignores someone working unsafely even once, credibility immediately slips. An effective field leader takes the offending crew member aside and makes it clear that production is not important enough for anyone to be hurt or killed. The field leader also explains how an accident can affect the crew member's family, both financially and emotionally, if the worst were to happen. Then the field leader reiterates the importance of commitment to safety and reminds the crew member that safety is a team effort.

If a field leader waits for someone on the crew to bring up all the safety concerns that need to be addressed, it may be a long wait. Crew members who know there is a problem, but choose not to express it, are generally concerned about the negative attention they may get from their field leader if they speak up. It is up to the field leader to assure the crew that this will not be the case and that their safety and well-being is a priority. An effective field leader encourages the crew to spot safety concerns and bring up issues. This message encourages crew members to speak up. The field leader is ultimately responsible for running a safe and productive job site.

Risky Behaviors

Despite safety rules, there will always be individuals who choose not to wear safety gear because they feel it is cumbersome or they do not see the need for it. **See Figure 4-2.** Any individual who is cavalier when it comes to safety, or who misunderstands what is being asked and does not ask for clarification, is more likely to make poor decisions. Workers get hurt when they think they are indestructible. An effective field leader deals with these individuals swiftly and effectively. Idle threats will not work.

Figure 4-2. Despite safety rules, there will always be individuals who choose not to wear safety gear because they feel it is cumbersome or they do not see the need for it.

The most common reason for an individual to risk jeopardizing safety is the fear of losing too much production time. To stifle this thinking, an effective field leader clearly and frequently communicates that safety comes first.

The field leader also shows the crew that this is true by taking every safety concern seriously. A field leader only needs to discount a safety concern once to discourage the crew from bringing up any safety issues in the future.

Problem Solving

There are also individuals who participate in unsafe practices out of fear of being ridiculed by other crew members. Ridicule is detrimental to a safe work environment, and the field leader needs to step in whenever necessary to put a stop to it. No one wants to be seen as the weakest link. This is why communicating expectations to the entire crew is so vital to their safety. An effective field leader needs to convince the crew that they are supported, but that they are also responsible for taking care of each other. It is a field leader's responsibility to have a one-on-one conversation with individuals who do not want to participate in a safe work environment and let them know there will be consequences following any unsafe actions.

Some people take unnecessary risks because they think they are "taking care of business" in a timely manner. These people are not only putting themselves in a compromising position but they are putting their field leader in one as well. Any time an accident occurs on a job site, the first person an investigator wants to speak with after talking to the injured party is the field leader.

Litigation

Being party to unsafe practices can land a field leader in court as a witness or even as a defendant. Death and injury on a construction site always mean someone will be looking to be compensated for what was lost. For an injured crew member, it might only be wages for lost time and medical expenses. If the injury causes the crew member to be permanently disabled, then the stakes can be as high as compensation for a lifetime of lost wages. When a family loses a loved one, there is only one type of compensation possible—money.

If a field leader is found to be responsible for an individual getting hurt or dying, that field leader will be front and center for as long as litigation continues. If it can be proven there was gross negligence on a field leader's part and that the accident could have been avoided, it may even lead to incarceration. An effective field leader uses sound judgment when considering the consequences of allowing crew members to work unsafely.

Leading in the Field: The Measuring Stick

The construction industry has always been a competitive work environment, and production is the measuring stick. When it comes to cutbacks and layoffs, every crew member knows that the highest skilled and most productive workers will be kept the longest. Their livelihood depends on their ability to stay on the payroll. A crew who is worried about slowing down production to incorporate safety measures is less likely to take those measures seriously. That is why it is up to the field leader to set the bar on safety and set it high.

No one should ever underestimate the guilt that comes with someone being injured or killed on a job site. Guilt forever reminds each person of their role in what occurred. It also asks, "What if?" time and time again. Anyone would be devastated if an accident happened because it was not made absolutely clear what was expected or because the proper safety equipment was not provided. An effective field leader does not allow inconvenience or a decrease in production time to be factors in a decision that may cause lifelong regret. Every crew member and field leader deserves to go home each night in the same condition they arrived in.

Section 4.1 For Discussion

1. Describe the role a field leader must take regarding safety.
2. What steps can be taken to improve productivity when following safety rules creates a perceived productivity issue?
3. Explain how crew members perceive the value of safety on the job site.
4. How do unsafe practices lead to unnecessary risks and potential litigation?

Section 4.1 Take Action

1. Draft a list of safety practices that a field leader could use to emphasize safety on the job site.
2. Role-play a field leader catching a crew member performing a task without using safety gear.

Section 4.2

DOCUMENTATION

Supervisors, project managers, and owners/customers often require documentation from a field leader throughout the length of a project. The size and type of the project impacts the amount of documentation required. Larger projects require a larger amount of documentation, while smaller projects cannot economically support extensive documentation. A field leader who fails to keep and organize project documentation will lose time and money. Keeping accurate records of change orders, daily job logs, RFIs, purchase orders, packing slips, coordination meeting action items, job progress, and as-builts (redlines) simplifies the compilation of a final report.

Leading in the Field: Detailed Documentation

A field leader is just finishing up a yearlong project that has had its share of challenges. But overall, it was a success. The project was completed on schedule, and the amount of labor hours used fell in line with the bid. Based on how the project was managed, the contractor asks the field leader to manage the next big job, which starts in a few weeks. The only thing the field leader has left to do is deliver all the required documents to the general contractor.

First on the list is putting together a set of comprehensive as-builts. The field leader digs through piles and piles of prints, but has trouble deciphering which marks most accurately describe the installations. Many of the prints had been shuffled around the job site since day one, and now there is no way of knowing who on the crew made the notes. The field leader makes a good guess while marking up the prints and hopes the customer will not notice the inaccuracies. Struggling to make sense of all the scribbles and scratches, the field leader thinks about how easy it would have been to make each redline mark on the very day an installation was altered.

Next, the field leader compiles all of the signed change orders that were authorized over the length of the job and makes sure the company has been or will be paid. The field leader knows that the money accumulated in the change orders is guaranteed to help the project reach a higher profit margin. After scrounging through the file cabinet for over an hour, and hoping they've all been found, the field leader realizes on closer inspection that two big change orders are missing. The field leader thinks back to a couple months earlier when the crew completed the changes, but has trouble remembering filling out the change order and getting it approved. The field leader dreads having to explain this to the project manager and hopes that the general contractor will still sign a change order for the extra work.

It is not uncommon for a field leader to lose focus on documentation when a project seems overwhelming. Effective field leaders understand the long-term consequences of ignoring proper documentation and use organizational systems on each project to get it done.

Change Orders

Change orders are used when work outside the scope of the construction documents is required. A change order is one of the most important documents that a field leader is responsible for. Not only will a field leader create and execute the change order but many times is also responsible for pricing, getting an approval signature from the owner/customer, and then tracking progress of the payment. One misstep in this process and the contractor may be left holding the bag after having provided the material and labor to make an installation without a clear path for being paid. It should go without saying that any unpaid change order has a direct negative impact on the bottom line.

As a field leader navigates a project, it is imperative that change orders are given the attention they deserve. Each change order must clearly state the work that is being added to the project; the cost of labor and materials; the identity and signature of the person(s) authorizing the change; the timeline in which the work will be completed; and any other specific and relevant information. All of these items must be in place and agreed to before any new work is done.

There may be times when a field leader will be requested to provide documentation for change orders, even after a project has been completed for some time. Although someone on the project management team may invoice and confirm payment of change orders, a field leader may be asked to provide supporting documentation due to a difference of opinion between the contractor and the owner/customer over what was approved and the work that was completed. No one but the field leader has had the hands-on involvement on the job site or can confirm what was said in conversations at the time the change order was requested.

Daily Job Logs

It is important for a field leader to document each day's unique project notes into a daily job log. **See Figure 4-3.** Items worth recording might include weather conditions; deliveries of materials; discovery of hidden site conditions; safety-related issues or incidents; and on-site visits by third parties such as a project architect, engineer, or owner/customer. The daily job log should also include a brief summary of conversations pertinent to the project that were had with on-site visitors. There are many different types of job logs a field leader may use. They can be a simple spiral notebook, a preprinted form provided by the contractor, or a digital file.

For example, a safety hazard is identified by a crew member, and the field leader communicates that information to the general contractor. Two weeks go by, and the general contractor does not take any action. Then an accident takes place, and the field leader's logs become part of the investigation into the cause of the accident. The field leader's daily job log documents every conversation, the people spoken to, and the dates and times when an attempt was made to have the safety hazard removed. The value of the information regarding this one event is solidified by the detail and care taken to document less important events in the previous days, weeks, and months leading up to the accident.

Daily Job Logs

Daily Job Log

Name: John Smith

Date: 5/10/20XX

Project: City Medical Office Building

Weather Conditions: Morning partly cloudy afternoon rain/62°F

Crew Size:	8
Journeymen:	6
Apprentices:	2
Helpers:	0

Site Conditions/Safety Concerns: Talked to the general contractor about loose handrail on the 6th floor. He said he would send someone up to fix it.

Work Being Completed: Wiring is being pulled from the MDP-2 to Panels E241A and E242A. Rough in of can lights in the 3rd floor hallway. Branch home runs are being spliced.

Special Visitors: The electrical engineer was on site today. Discussed the two outstanding RFIs we have not received answers on. He said that we should see them by the end of the week.

Deliveries: Ace Electrical Supply made a delivery today that wasn't fully complete. This has been an ongoing problem. I called and talked to Bill Thomason and explained how these mistakes were costing us time and money. He said he was unaware of the previous instances, but would get to the bottom of it and get it corrected.

Coordination With Other Trades: Laid out the penthouse mechanical room with the pipefitters today to make sure we all have the clearances we need.

Comments: Jeff was late again for work today. I talked to him about it and let him know that I would have to write him up the next time. He apologized and said that it wouldn't happen anymore.

Figure 4-3. It is important for a field leader to document each day's unique project notes into a daily job log.

Job logs can become tedious and seem unimportant compared to other daily responsibilities a field leader has. However, it is the field leader's responsibility to protect the interests of both the crew and the company, and an updated daily job log is one way to accomplish that.

Requests for Information (RFIs)

No matter how diligent the architect and engineering teams are when putting construction documents together, a field leader is always going to find discrepancies and missing information that needs to be clarified with an RFI (request for information). A field leader's job is to identify those discrepancies and missing items as early as possible and get clear direction on how to move forward. Once RFIs are answered, it is important that they are stored in such a manner that a field leader can find and reference them easily.

The field leader of every trade on a project compiles a list of RFIs. Some of the questions asked will affect another trade's installation and some will not. As responses to RFIs pour in, it is common for a field leader to become overwhelmed with the task of keeping them all organized. An effective field leader establishes a system for keeping RFIs organized before they pile up. **See Figure 4-4.**

RFI Log

RFI Log

Company: Medford Construction Services

Project Name: City Medical Office Building

RFI NO.	Description	Trade	Date Written	Date Returned	Comments/Impact
001	Clarify fixture type in Room 226	Electrical	03/05	03/13	No Impact
002	Mechanical Room Layout Equipment Conflict	Mechanical	03/07	03/13	No impact
003	Clarification of Wall Type C	Framing	03/09	03/14	Same as Wall Type B except using 6" Studs
004	Soffit framing dimensions in main floor lobby	Framing	03/10	03/15	Reference RFI for new drawing
005	Clarify Flush mount or surface panel	Electrical	03/13	03/17	Write change order for new framing requirements
006	Door swing in room 201	Framing	03/14	03/21	Order new door swing opposite of drawing
007	Missing dimension on Sheet A604	Framing	03/15	03/21	See RFI
008					

Figure 4-4. An effective field leader establishes a system for keeping RFIs organized before they pile up.

Another reason an organized filing system is needed is that many RFIs written at the beginning of a project will not affect an installation until many months later. It can be extremely frustrating for a field leader to know that an RFI was written and responded to and yet not be able to figure out what number it was or where it was put. And architects and engineers do not take kindly to being bothered with the same question more than once on a project.

Purchase Orders and Packing Slips

The documentation of purchase orders and packing slips is a field leadership responsibility that often gets pushed to the bottom rung of importance, but it should not. Larger purchase orders for big-ticket items are fairly easy to track and account for. But it is the smaller day-to-day orders that can end up causing a field leader big problems. Over the length of a project, a field leader will write more small purchase orders than large ones.

Small purchase orders are written to many different vendors for a variety of materials and services. As deliveries come in, they must be checked for accuracy, which means comparing the packing slip that the vendor provides against the purchase order that the field leader submitted. This is no easy task but one worth attention. Once a delivery is made and has been signed for, the burden of proof falls on the field leader if a vendor did not deliver everything listed on the packing slip.

Purchase orders may also be double billed, and vendors may charge for items that were not ordered or received. Whether or not these errors are accidental is irrelevant. What is important is that the field leader has a system for checking in deliveries and then filing, storing, and referencing all purchase orders and packing slips related to the project.

Coordination Meeting Action Items

Taking notes during coordination meetings and keeping the notes organized is a great way to record what is said by others and the commitments they make to production schedules. Many general contractors will write down all of the meeting notes and action items and then distribute them to all of the trades at the conclusion of each coordination meeting. This is helpful because it is another reference that may come in handy. However, nothing can replace the notes a field leader takes. An effective field leader gets needed answers from the field leaders of other trades and then thoroughly documents that information.

Job Progress

Job progress is often documented through the use of job journals, but pictures or short videos are a valuable supplement. Time-stamped photos of actual job-site conditions can be useful when it comes to documenting job progress. It is easy for a field leader to take out a cell phone or tablet and shoot some photos each day to document job progress. **See Figure 4-5.** Most of the photos should document what the crew has completed. Other reasons a field leader may take photos is to document safety hazards that need reporting and the progress, or lack of it, made by other trades—especially when it directly affects the crew's ability to complete an installation.

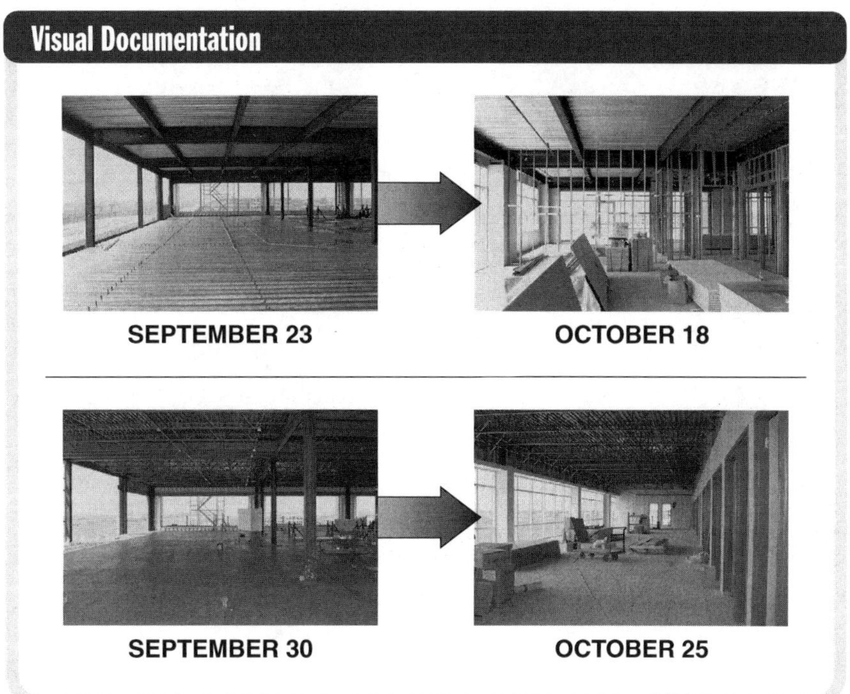

Figure 4-5. It is easy for a field leader to take out a phone or tablet and shoot some photos each day to document job progress.

As-Builts

Field leaders use as-builts (also known as redlines) to record any changes made to original construction drawings in the field. Spelled out very clearly in every construction contract is a requirement for each contractor to provide one or more full sets of as-builts once the project has been completed. One way a field leader can document all the changes that occur as they happen, and keep them in one place, is to have a full set of prints dedicated to nothing more than being marked up to track as-builts. It is common for a field leader to have multiple sets of prints in use on a single job site—a field set for the crew to reference, a personal working set, and a set dedicated to as-builts. An effective field leader tries to avoid waiting until the end of the project to mark up as-builts. Digging through multiple sets of prints to compile random notes that are scribbled across every single drawing that was used throughout the project is not an efficient or accurate way to approach this responsibility. An effective field leader makes updating as-builts a daily priority.

Section 4.2 For Discussion

1. How does the size of a project impact the amount of documentation required?
2. Identify the types of information that must be clearly stated in change orders.
3. Explain how the details recorded in daily job logs are used for documentation.
4. How are RFIs, purchase orders, and packing slips used as documentation?
5. How do coordination meeting action items document the commitments made by the various trades on a project?
6. How does adding visuals to job journals help document a project?
7. Explain the value of dedicating a set of prints to the tracking of as-builts (redlines).

Section 4.2 Take Action

1. Complete a change order and explain how it will be documented in the final project report.
2. Take photos of job progress and then integrate the images into a job journal.
3. Exchange tips for allocating adequate time for documentation.

Section 4.3

PRODUCTION CHALLENGES

Each day of production, there are orders that need to be placed, deliveries that need to be checked in, and coordination issues to resolve. It is impossible for a field leader to perform these tasks without the assistance of a qualified crew. Delegating is the act of assigning a task to another person while remaining responsible for the results. To delegate tasks, field leaders must clearly communicate the expected results to the crew members assigned to each task. Monitoring a crew's progress increases the possibility that tasks will be finished according to schedule.

An effective field leader must always be on the lookout for new ideas and methods that will make the job easier and improve the chances for success. The day a field leader knows it all, or becomes content with doing just enough to get by, is the day that field leader will be bypassed. Following are challenges that can hurt production when a field leader is not doing an effective job.

Lack of Planning

An effective field leader knows that the key to managing an efficient job is always having what the crew needs when they need it. There is only one sure way to know this is happening and that is by spending the appropriate amount of time creating a detailed project plan. This plan must account for writing any needed RFIs, ordering materials, and having additional tools or materials brought to the job site. The result of bypassing this integral planning step is guaranteed to have a negative impact on both crew efficiency and production. Every time a crew member approaches the field leader looking for a new task because of a lack of what is needed to do the assigned task results in wasted labor hours and loss of profit.

Leading in the Field: New Crew Members

A field leader is running a project that requires more crew members, so the field leader contacts the project manager to request more manpower. The next day, new crew members show up, and the field leader gives them a brief orientation to the job site and sends them off to work.

The new crew members end up spending most of their first day trying to get tools together to get started. By day's end, they are frustrated with not being able to be productive right away. The field leader did not take the time to put together what they needed in advance or even confirm that what they needed was on site. If the field leader had thought about it in advance, an apprentice could have been assigned the task of rounding up everything the new crew members needed ahead of time. If they were short on anything, it would have given the field leader at least a day's notice to order those things, and the new crew members would have been able to go right to work.

Lack of Tools and Materials

It eats into profits when a field leader does not provide the crew with the appropriate amount of tools or materials. Every decision a field leader makes regarding production directly impacts the bottom line. If crew members have to look for tools and materials that should be readily available, their field leader needs to recognize the problem and quickly find a solution. **See Figure 4-6.** Every hour a field leader ignores this issue, valuable labor hours are wasted.

Misplaced Tools

Figure 4-6. If crew members have to look for tools and materials that should be readily available, their field leader needs to recognize the problem and quickly find a solution.

When a crew member is standing with tools in hand and looking uncertain, the field leader must stop to assess the situation. For example, the crew member may be ready to install lights in a 12-ft ceiling, but not able to find a tall enough ladder anywhere. An effective field leader knows that even if a ladder is borrowed from another trade, it is likely it will have to be returned before the installation is complete. This oversight means that the field leader must borrow a ladder with the hope that the crew member will not be interrupted in the middle of the installation or reassign the crew member to a new task while a ladder is delivered. Both of these options have disadvantages that are sure to cause a downturn in efficiency and production. But the truth is proper planning would have avoided this altogether.

To offer another example, a crew is building hundreds of pipe racks made of strut and threaded rod that will support thousands of feet of conduit. Three days into the project, the crew runs out of nuts, washers, and other inexpensive materials. Production comes to a halt. Yes, the field leader may be able to borrow some from another trade temporarily but never should have run out of inexpensive items like those. A field leader would have to buy a lot of nuts and bolts to make up for the wasted labor hours that accrue by not having sufficient quantities in stock.

A crew can have the best intentions for completing a project in a timely manner, but their field leader cannot expect them to work magic if what they need is not provided. A lack of tools or materials hits a field leader right where it hurts the most—in wasted labor hours.

Lack of Organization

When it comes to organization, the field leader sets the example for the crew. Keeping the prints, the trailer, or the work truck in shambles tells the crew that organization is not one of their field leader's priorities. If the crew sees that their field leader does not mind working in a disorganized setting, then they will not stay organized.

Setting Clear Expectations

Likewise, crew members should always be able to find the tools and materials they need within a reasonable amount of time. If members of a crew have to go to three different locations to find all the different materials they need to complete a task, labor hours are wasted. And if multiple locations are used to store tools and materials, an effective field leader makes sure that similar items are stored together and informs the crew if these items are moved to a different location.

Leading in the Field: The Missing Drill

A field leader has at least one crew member who spends half an hour a day searching for a drill. If each crew member has a burn rate of $50 an hour, at half an hour a day, that is $125 a week wasted while a crew member searches for a drill. This means that after three weeks of the crew being short just one drill, the project has taken a $375 loss in labor. That amount of money could have purchased a brand new drill and eliminated the problem altogether. But either way, $375 has come off the bottom line of the project.

The difference is that once a drill is purchased, the need to search for it ceases to exist, and no more labor would be wasted looking for it. If the field leader decided not to purchase a drill, there is no way of knowing what the final tally of wasted labor hours would be by the end of the project.

This example could be applied to any tool or material that a crew might not have on hand. An effective field leader understands that the cost of labor far outweighs the cost of most tools and materials and makes sure the crew has what they need to do their jobs.

Lack of Coordination

It takes a lot less time to coordinate with other trades and talk through a plan that works for everyone than it does to take something apart and rebuild it once it has already been installed. Failing to coordinate efforts or choosing not to look at the other trades' prints to see what they have planned only increases the likelihood that there will be some significant rework coming the crew's way.

Poor People Skills

A field leader has to be able to put individuals in positions that best suit them but that will also work to the project's advantage. Once a field leader learns to identify the strengths and weaknesses of each crew member, this information can be used to improve their performance as well as the project's outcome. For example, if a field leader asks a crew member to do something the crew member may only have done a couple of times before, the crew member may feel unsure but say nothing. The crew member's work will progress slowly and may result in an unsatisfactory installation. Had the field leader taken the time to ask a few questions about the crew member's experience prior to assigning the task, it would have been apparent that the task should have been assigned to another crew member.

Poor Communication

A field leader's ability to communicate information to the crew is crucial to project success. Many field leaders hurry through the layout of a project or leave out valuable details only to end up frustrated when the crew does not do what the field leader expects. Every detail that a field leader leaves out is one more detail left up to the interpretation of the crew. The only way for a field leader's expectations to be met is for them to be clearly explained.

The only way for a field leader's expectations to be met is for them to be clearly explained.

An effective field leader spends a lot of time studying prints and dissecting a project piece-by-piece. The crew should not be expected to have anywhere near the knowledge of the project that the field leader does. Some field leaders choose to not make prints available to their crews, and some field leaders dole out small portions of the prints over time. Both are a huge mistake. Effective field leaders give their crews access to the prints, which allows them to double-check details and find answers to questions that arise as they work. This approach saves a field leader time and helps eliminate preventable mistakes.

Poor Crew Morale

Poor morale is a production killer that can easily go unnoticed. Poor morale often stems from the lack of attention to communication, organization, and the appropriate provision of tools and materials. Unfortunately, a disgruntled crew often does not communicate discontent with their field leader. Instead, they commiserate among themselves, compounding the problem and often dragging crew members who were otherwise content into the fray. An effective field leader needs to recognize the signs of poor morale and come up with solutions to turn it around.

Crew Inconsistency

A field leader is constantly required to switch gears and change direction throughout a project. **See Figure 4-7.** This makes it difficult to find a consistent path for completing each phase of a project from start to finish. During these transitions, many different people often work on a single phase of the job. Having one crew member pick up where someone else left off creates the opportunity for details to be missed. That means it is up to the field leader to fully communicate the nature of the work originally given to the person who started that phase of the project. Otherwise, a step may get skipped, a mistake may be made, or a loose end created.

Changing Direction

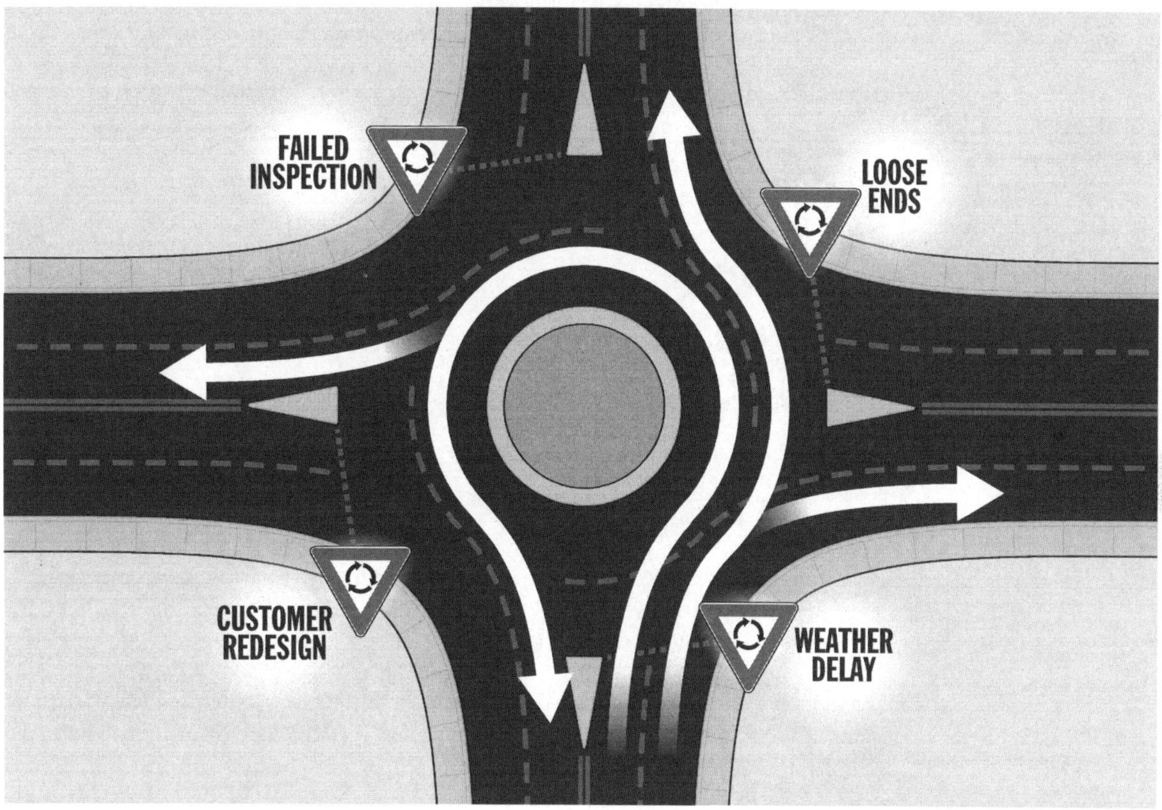

Figure 4-7. A field leader is constantly required to switch gears and change direction throughout a project.

Construction projects also go through cycles of being busy and slow. This means that crew members who work with a field leader at the beginning of a project may not be there at the end of the project. Some crew members may leave for another project when they are no longer needed. When a field leader wants them back, it may be difficult, if not impossible, to get them to return due to timing. It is best to have a crew who is around for most of the project. It is an obvious advantage to a field leader to have a consistent crew throughout a project because they will much more likely see the "big picture." Bringing in crew members near the end of a project who are unfamiliar with it does not lead to optimum production.

When it is necessary to bring new crew members into an already productive job site, an effective field leader attacks this production challenge by having clear, concise projects for them to complete. The field leader also has all the necessary tools and materials set up exactly where they are needed—in the work area—and has the newcomers work with or around someone who has been on the project long-term so it is easier for their questions to be answered.

When a project is in full swing and new crew members arrive on site, existing crew members can become frustrated because answering questions slows down their production. To address this, an effective field leader identifies a current crew member working in the area as the "go-to person." The field leader's acknowledgment of the current crew member's worth is often enough for the crew member to welcome the added responsibility. However, there are crew members who may expect additional compensation for taking on the added responsibility, so this may not be an option.

Avoidable Mistakes and Rework

Mistakes and rework go hand in hand. That is why incorrect installations have such a negative impact on the bottom line. Any time a field leader has someone take something apart in order to reinstall it, the labor hours spent go directly into the red column.

Sometimes mistakes do not become apparent until later in the project when they are much more difficult to correct. Of course, not all mistakes are avoidable. An effective field leader stays focused and plans well to keep mistakes and rework to an absolute minimum.

Leading in the Field: Long Leads

An order for a long-lead item is placed incorrectly. The field leader recognizes that it was an avoidable mistake. Unfortunately, fixing the problem is not easy because the order was for a custom item made specifically for the project. So the field leader has no choice but to reorder the item.

In addition to not being able to finish the project, the field leader has to explain the situation to the customer and take responsibility for the cost of the incorrectly ordered item. If the field leader had spent a few extra minutes going over the order, or had someone else review it before it was sent in, then the costly mistake may have been prevented.

The intended use of a building influences the materials and installation methods that can be used. An effective field leader never assumes that a project is going to be built exactly like the one before. Buildings like hospitals often require unique installation methods and have special code requirements. These projects need to be thoroughly researched by the field leader so they can be correctly installed. If a crew undertakes an installation based on assumptions rather than the required specifications or code regulations, the field leader could end up having to rework the installation. Unfortunately, rework is a common occurrence.

Loose Ends

The nature of construction is to work on different parts of a project in phases, starting and stopping multiple times to allow other trades to do their work. Having too many loose ends at the end of a project can be a frustrating and pressure-filled experience for a field leader. It is the kind of pressure no one needs or wants. Solid planning and consistent note-taking can guide a field leader through a project in a way that avoids many potential loose ends.

When loose ends slip through the cracks, a field leader will end up with a profit killer in the end. Any extra hours a field leader may have saved through efficiency and production slowly slip away as the crew spends hours trying to first identify the loose ends and then finish them. This might be as simple as making a label or it may involve ordering more materials that the field leader was sure they were not going to need.

It is the field leader's responsibility to make sure that loose ends do not pile up. The key to this is making sure that everything that can be finished is finished. If a field leader pulls somebody off a project before it is finished, a thorough record should be kept of what remains to be done. Then as soon as possible, the field leader needs to send a crew member back to tie up every loose end. An effective field leader also lets the crew know that their help is needed and asks for updates about any loose ends left behind when they step away from a task that they may have otherwise completed.

Section 4.3 — For Discussion

1. Who is responsible for tasks that have been delegated?
2. How does a lack of planning, tools and materials, organization, and coordination lower production efficiency?
3. Explain how poor people skills, poor communication, and poor crew morale lower production efficiency.
4. How does crew inconsistency lead to avoidable mistakes, rework, and loose ends?
5. What can a field leader do to make sure loose ends do not kill the profit on the project?

Section 4.3 — Take Action

1. Outline a plan that a field leader can follow to address common production challenges.
2. Role-play two field leaders from different trades discussing how they address production challenges.
3. Role-play a field leader asking a seasoned crew member to bring a new crew member up to speed on a project that is two weeks from completion.

Section 4.4

THE STRESS FACTOR

Stress is serious business. The construction industry is a very stressful environment, and those in field leadership positions experience stress tenfold. Many factors in a construction project can lead to stress. **See Figure 4-8.** Field leaders affected by stress may have trouble sleeping, experience strained personal relationships, or have to cope with stress-related illnesses that deteriorate their health.

Factors Contributing to Stress

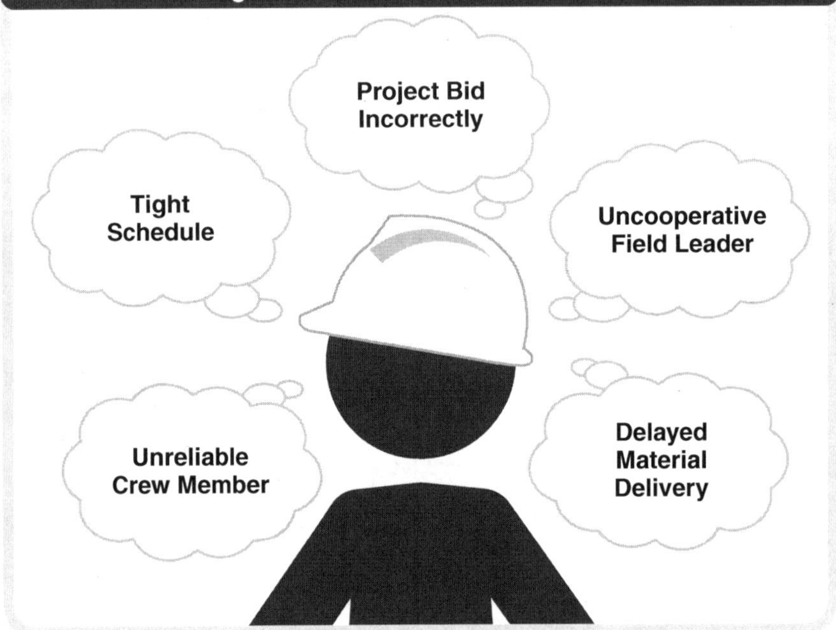

Figure 4-8. Many factors in a construction project can lead to stress.

It is vital that field leaders maintain their perspective about what is most important. A lack of communication between the field leader and the crew can directly affect a field leader's stress level as well as the outcome of a project. Yes, the job is important and so are commitments to owners and customers. But if a field leader does not have good health, none of it will matter in the long run. If a field leader deals with stress by abusing drugs or alcohol, has a nervous breakdown, or has a heart attack, a replacement will be found. It is a harsh reality, but the projects the field leader is running will go on. Taking stress seriously is the first step toward controlling its effect.

Stress is unavoidable, but it is manageable. The first step to managing stress is to recognize that some is preventable and some is not. Learning to cope with the pressures that come with running work is an absolute must if a field leader wants to avoid having stress 24 hours a day. An effective field leader does not waste time or energy trying to control things that cannot be controlled but focuses on making strides toward relieving pressures that come with the job. An effective field leader continually monitors stress levels and takes measures to prevent the stress from being overwhelming. Managing stress is a must for those who wish to have long, productive, and rewarding careers in the construction industry.

Identifying Stress

The best way for a field leader to identify stress is to step back, analyze a typical day, and then pinpoint when stress occurs most. A field leader determines if the stress is being caused by a lack of planning or a lack of communication. If a field leader views responsibilities that take a lot of time and focus, such as studying prints and preparing plans, as a source of stress, then there is only one thing to do. No one else can perform these tasks, so an effective field leader sets aside time each day when interruptions will be at a minimum and gets these tasks done.

A field leader determines if the stress is being caused by a lack of planning or a lack of communication.

Most causes of a field leader's stress have reasonable solutions. Sometimes those solutions take a while to find, and sometimes they come from unlikely places. The key is to avoid believing that problems cannot be fixed. A defeatist attitude will only cause more stress. Three primary causes of stress are unpreparedness, ineffective scheduling, and unpredictable events.

Unpreparedness. A lack of preparedness is one of the most preventable forms of stress that a field leader will encounter. No field leader wants to be considered unprepared or out of control by the crew or by field leaders of other trades. If this happens, considerable hostility may be directed toward the field leader, which will manifest itself in the form of both mental and physical stress.

An effective field leader does not arrive at meetings without having put effort into a plan that spells out the execution of an upcoming phase of the project. A field leader must take the time to study the drawings of the other trades and have a basic understanding of their installations.

An effective field leader does not get angry when asked a question, nor feel embarrassed if an answer cannot be given right away. There is no possible way a field leader can know the answer to every question. However, when faced with the reality that not enough tools were ordered, a field leader will likely become irritated or angry. If a field leader takes responsibility when a mistake is made, stress can kept at a minimum.

Ineffective Scheduling. Project schedules help ensure the appropriate number of crew members are scheduled for each phase of the project. Having too many crew members scheduled increases labor expenses, which cuts into the bottom line. And scheduling fewer crew members than needed can yield poor results. If scheduling is the cause of stress, the field leader should ask the following questions:

- Does the crew have the opportunity to be as efficient and productive as possible?
- Have realistic expectations been set regarding the capability of the current number of crew members to complete the tasks scheduled?
- When asked a question, is it hard to say, "I don't know, but let me do some research and I'll get back to you with an answer"?

If "yes" is the answer to these questions, the field leader should ask the crew if there is anything that can be done to boost their level of production. Crew members have a different perspective and often have ideas for solutions. If crew members do not have feasible solutions, then the field leader should talk to the project manager rather than give up. The further behind schedule the crew falls, the more stressed the field leader will become and the less likely they will complete the project on time and within budget.

Unpredictable Events. Construction work by its very nature can be unpredictable. Deliveries do not always arrive when expected; the weather often does not cooperate; and a trade may not pass an inspection causing a shift in another trade's schedule. Countless variables are out of a field leader's control. That is why it is important for a field leader to have a contingency plan as a safety net. Whether this plan ends up being implemented or not, it can help lower stress. Detailed advanced planning also increases efficiency and production. Having a long list of tasks that crew members can accomplish while the field leader deals with unanticipated circumstances can also lower a field leader's stress. Since no plan is etched in stone, a field leader must also be cooperative and flexible in order for all of the trades on a project to succeed.

An effective field leader takes ownership of things that can be controlled and handles the rest in a way that keeps the project productive and on schedule.

Unpredictable stress caused by unplanned events can also cause a field leader to make excuses and place blame, which undermines a field leader's credibility. An effective field leader does not point fingers when things do not go as planned but takes responsibility and finds solutions. An effective field leader takes ownership of things that can be controlled and handles the rest in a way that keeps the project productive and on schedule.

Lowering Stress

There are days when it seems like every person who crosses a field leader's path is demanding something right now! While each person probably has a legitimate question or needs information, it may seem like things are spiraling out of control at times like these. The best thing for a field leader to do is to address one question or request at a time. An effective field leader determines what is most important and then writes each item down in order of importance. Doing this only takes a few minutes, and having a written list provides focus. As each issue is addressed, the field leader's stress level will drop.

Stress levels can also be lowered by the crew informing the field leader in advance of when they expect to finish each task. This advance notice gives a field leader time to properly assess what direction the crew member should take next, which will lower the field leader's stress level.

Another way a field leader can lower stress levels is to avoid tackling too many tasks at once. For example, when a field leader knows that an RFI needs to be written, a phone call returned, an email written, and the crew checked on, a decision must be made as to which task needs to be done first. Making that decision is not always easy, but it must be done. An effective field leader does one thing at a time and does it well.

A field leader who becomes paralyzed by all that needs to be done and does not make decisions wastes precious time. Sometimes a field leader needs to ask for help. Unfortunately, many field leaders think that asking for help means admitting incompetence. While every individual should take pride in work well done, a field leader should not be concerned about asking for help. An effective field leader understands the value of learning from the experience and expertise of others.

Reasonable Expectations

A field leader is entitled to have a list of reasonable expectations that need to be met so that the crew can do the best job possible. This might include the management team holding a prejob meeting to discuss the upcoming project. This prejob meeting is when a field leader should ask questions that are pertinent to how the project will be approached. Topics of discussion may include the following:

- How much is the contract amount?
- How many labor hours were bid into the project?
- Where do the project managers' responsibilities end and the field leaders' begin?

Answers to these types of questions are important. Without clearly defining the parameters of what a field leader is responsible for, things can be overlooked. And when things get overlooked on a construction project, contractors lose money. And when contractors lose money, they want to know who is responsible. An effective field leader manages stress by clearly understanding what is being asked.

> An effective field leader manages stress by clearly understanding what is being asked.

The prejob meeting is crucial to any project in terms of the future dynamics and relationships of the team throughout the length of the job. A field leader cannot assume that everyone will provide all the needed information. Getting answers during the prejob meeting is the field leader's chance to make sure that support from the team that put the project together during the bidding phase is there.

A field leader should always take the opportunity to ask questions about a project. Even if it does not happen as a sit-down meeting with the entire management team, in-depth conversations with anyone who has been or will be involved with the project are helpful. For example, project managers and estimators make determinations about the building process during the bidding phase. They can provide pertinent information such as details about the materials to be used, the vendors chosen to supply the materials, and installation methods.

At the conclusion of a project, especially if it has lost money, a field leader will more than likely be asked to explain how and why the project failed to make a profit. An effective field leader brings all the documentation from prejob conversations and any other pertinent information that will prove that, given all the circumstances, the best job possible was done. The field leader does not make any excuses. If there are things that need to be done better, the effective field leader owns up to them and also explains what was learned from the experience and how this knowledge will increase the probability of the field leader's success on future projects.

Self-Preservation

An effective field leader needs to be physically, mentally, and emotionally healthy. This includes making timely decisions regarding boundaries. It also encompasses a willingness to stand up for what is right and an ability to make tough decisions. Some of the decisions a field leader makes will be seemingly inconsequential, but other decisions can affect the field leader's employability and the level of career success achieved. But if a field leader has exceptional trade and leadership skills, opportunities will arise.

Each time an effective field leader takes on a project, large or small, questions must be asked. An effective field leader does not let ego, pride, desperation, or disillusionment give way to poor decision making. There will be times when an estimation team is not completely forthcoming with information, such as an underbid project. This lack of full disclosure does not stop with the estimation team because they still have to find a field leader willing to run the underbid project.

Leading in the Field: An Underbid Project

A field leader is asked to run a project for a contractor that the field leader has worked for many times before. This project was bid incorrectly, but it is impossible for the contractor to get out of building the project and still maintain credibility with the customer. If the contractor were to bail on the contract, the customer could choose not to invite that contractor to bid on any future projects.

Should the field leader agree to take the project? An effective field leader who is willing to run difficult projects that are not destined to turn out as profitable as others can become a valuable asset to a contractor. The key for any field leader stepping into this position is to have an honest and open dialogue with the contractor/owner about what the realistic goals of the project are and what the best and worst outcomes might be.

It is likely that at some point, a field leader will be asked to run a project that has been underbid. In the spirit of being a team player, the field leader should discuss the entire project at length with the management team, including the best- and worst-case scenarios for the project's completion. Is breaking even an acceptable outcome? What if the project comes in 3% over the prejob expected outcome? Will repercussions be levied against the field leader if the job goes completely upside down? An effective field leader will make sure all those questions are answered before taking on a losing job. If the questions are not answered, the field leader should decline the position.

On the other hand, if the field leader is hired to mitigate potential loss, accepting the position could still have a positive outcome for the field leader's career and relationship with that contractor/owner. If the project goes sideways and people are scrambling to protect their hides, not everyone will be willing to back up what the field leader says or does. An effective field leader documents every prejob meeting and conversation that takes place and the conclusions that were reached prior to taking on a project that has been underbid.

When a field leader thinks that an opportunity may be taken away because of too many questions or concerns or that something is not quite right with a project, an effective field leader "uses the brakes" and walks away from those opportunities, knowing that they could be career-damaging.

Stepping Up

Each field leader must decide when to tackle projects that are a challenge. Sometimes project managers play field leaders against each other in order to push them to excel. An effective field leader sees this behavior for what it is and does not play into it. There will always be field leaders who have more experience and skill. Instead of trying to outdo these field leaders, an effective field leader should try to learn more efficient job methods from these field leaders.

> Instead of trying to outdo other field leaders, an effective field leader should try to learn more efficient job methods from other field leaders.

Many field leaders are in such a big hurry to reach their career goals that they make ill-advised decisions regarding projects that may be beyond their expertise. Everyone wants a field leader who is confident, but being underqualified to run certain types of projects should not affect a field leader's ego because everyone has strengths and weaknesses, even after years of experience. Factors determining whether or not a field leader is qualified to manage a job include past experiences. For example, a field leader may have considerable experience working on industrial plants but may not have worked on high-tech manufacturing facilities. These two types of projects are worlds apart in how they need to be managed and built.

The people who decide who should run a project may not always consider if a field leader is ready to move to the next level. It is easy to spot overeager field leaders who have been less than forthcoming about their abilities because they are obviously overwhelmed by the breadth and complexity of a project. If a field leader wants to take on a more challenging project, a conversation with the contractor is in order. This does not mean that the field leader will end up running the project. However, it does give the contractor an opportunity to point out areas where the field leader could improve and to suggest job-shadowing another field leader to gain more experience.

An effective field leader advances at a pace that feels comfortable and does not let anyone insist on a course of action that is damaging in terms of reputation. Every field leader has an entire career to run larger and higher profile projects but only after becoming as effective a field leader as possible.

Work-Life Balance

If a field leader becomes obsessed with career progress to the point where it takes up time and energy on the job, it will cost in terms of personal life. There are times when a field leader has to say "no" or simply "not right now." A smart project manager knows that if a field leader is mentally exhausted, the work will suffer.

Working under extreme pressure and stress requires downtime. A contractor who respects and appreciates what a field leader brings to the table on an ongoing basis understands the need for breaks. These necessary breaks enable an effective field leader to come back to the job refreshed and able to function at highest capacity.

Effective field leaders balance their personal and work lives to avoid burnout. Focusing on work at the expense of personal life means little time and energy for family, friends, and recreation. An effective field leader works toward maintaining a healthy work-life balance. **See Figure 4-9.**

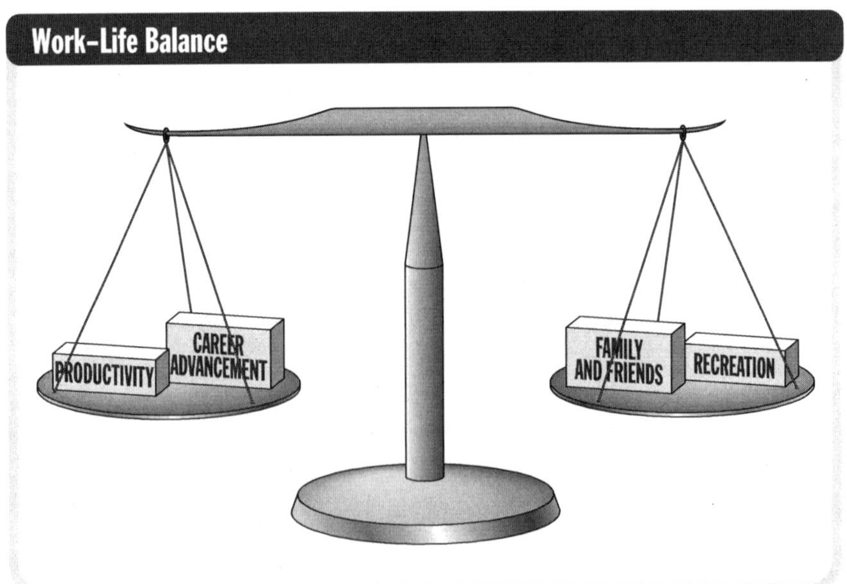

Figure 4-9. An effective field leader works toward maintaining a healthy work-life balance.

Section 4.4 For Discussion

1. Describe how unpreparedness, ineffective scheduling, and unpredictable events cause stress.
2. Why is doing one thing at a time the key to lowering stress?
3. Explain how obtaining information at a prejob meeting can help a field leader set reasonable expectations.
4. How does making timely decisions, making tough decisions, and asking questions help a field leader in terms of "self"?
5. Contrast the desire to step up to more challenging projects with the desire to maintain a work-life balance.

Section 4.4 Take Action

1. Create a table listing the causes of stress and potential coping strategies for each cause.
2. Role-play a field leader and a project manager discussing a stressful job-site situation.

Project Coordination

A new multiuse high-rise is being built in the heart of a downtown area. On the uppermost floors, ironworkers are nearing completion. On the lower floors, the framers, drywallers, plumbers, and electricians are making installations for their respective trades. Managing Simpson Electric's installation are two field leaders who are working together for the first time, Mike Jennings and Gary Hanson.

Mike has managed crews for over 10 years and believes solid planning, clear communication, and detailed documentation are the best ways to ensure a profitable job. Gary is working as a field leader for the first time. His approach to managing his crew is to order materials and tools as needed so that there is very little waste at the end of the project.

Mike and his crew of six journeymen electricians and three apprentices are responsible for power distribution. They have completed setting the main switchgear in the basement and are now working to run feeder conduits to subpanels on each floor.

Gary and his crew of seven journeymen electricians and two apprentices are responsible for branch power throughout the structure. They have roughed in the first two floors and are now beginning work on the third. Gary's crew is also responsible for running conduit in the concrete slabs before each pour.

Monday **8:00 am**

Most of Gary's crew is picking up where they left off Friday afternoon roughing in the third floor, but he is sending one journeyman and one apprentice to the ninth floor to run conduit on the deck before the concrete is poured. The apprentice has helped with slab work before, but this is the first time the journeyman has performed this task on a project.

Gary:	"I need you two to go up to the ninth floor right away and get some conduit thrown in the slab. Here are a couple of sketches I threw together."
Journeyman:	"Is all of the material up there already?"
Gary:	"Most of it should be on the fifth floor. But if you don't find everything there, you might be able to scrounge up the rest of it in the basement."
Journeyman:	"I see on your sketch that some of these stub-ups are near the outside edge of the building. Are there tie-off points established that we can use? And, I can't remember the last time I saw a safety harness on this job let alone a retractable lanyard."
Gary:	"Look! They're planning to pour the slab first thing tomorrow morning, so we need to get moving on this right away, and I really need to tend to other issues right now! I need you to do what you are getting paid to do! Your apprentice should be able to help. He worked on a couple of the other decks."
Journeyman:	"Alright. . . I'll do what I can."

After this exchange, Gary heads to the third floor and is met by one of his crew members who reports that he is short an 8-ft ladder. Gary tells a journeyman to go down to the basement and borrow one from Mike's crew. The journeyman replies that the last time he went to the basement to borrow a tool, Mike jumped on his case. Gary tells him to pay no attention to Mike. Since they all work for the same company, all the tools are to be shared.

Case Study ...continued...

Monday **9:34 am**

Gary is approached by another crew member who reports that a plumber just told him that the pipe run he is working on is at the wrong elevation. The plumber insists that he take down the rack of conduit he is building because it is blocking their progress. Gary calls the plumbing field leader and asks to meet him at the location of the conflict.

Gary:	"I understand it would be nice for you if we moved the rack so your job would be easier. But the rack is installed already, so just work around it!"
Plumbing Field Leader:	"It's not about what would be nice for us. It's about what's in the RFI that I wrote three months ago discussing this conflict. Did you read it?"
Gary:	"There are a lot of RFIs to keep track of, and I'm not a plumber. Why would I read your RFIs?"
Plumbing Field Leader:	"Well, I hate to break it to you, but you're not going to win this one. I suggest you take this down right away so my crew can get back to work. I don't want to have to back-charge your company for holding us up!"

After this exchange, Gary makes his way to the job trailer to see if he can get some support from the general foreman.

Gary:	"Hey, you got a second?"
General Foreman:	"Sure. What's up?"
Gary:	"Do you remember seeing an RFI that the plumber wrote about an elevation change on the third floor?"
General Foreman:	"Yeah, that rings a bell. But I don't remember the particulars of it off the top of my head. Why do you ask?"
Gary:	"Well, we already have our pipe rack built, and now the plumber is demanding that we take it down because we're in his way. Do you know how much time that's going to take to redo all that work?"
General Foreman:	"You think I don't? Unfortunately, we don't have a choice in the matter. He wrote an RFI, and a decision was made to alter the way the prints read. Frankly, I'm a little frustrated that we're in this position in the first place, because I've talked to you about your lack of attention to RFIs. This is precisely why we keep an RFI log that includes RFIs from other trades. Sometimes the changes affect us, and sometimes they don't. I guarantee that if you look in the log, you'll find a comment regarding how his RFI affects our approach to that pipe rack."

Mike Jennings then enters the trailer.

Case Study ...continued...

Monday 9:34 am

Mike: "There you are! We need to get something straight! I don't mind loaning you and your crew tools once in a while, but this is getting a little ridiculous. I just ran into one of your crew members taking a ladder from the basement. He never even asked if we were going to be using it! On top of that, when I asked him where he was taking it, he said it was none of my business and that the tools belong to the company and not me. He said you told him to tell me that."

Gary: "I didn't tell him to say that. But it is true that we're all on the same team and we should be willing to help each other out."

Mike: "That argument of yours is getting old. Here's the way I see it. You don't plan ahead, and you don't get everything your crew needs, so you think you can take whatever you want without asking! How is that being a team player?! Have you ever stopped to consider that my crew never borrows anything from you?!"

General Foreman: "I'm with Mike on this. When you're a field leader, you have to make sure that your crew has all the tools, materials, and information they need to complete assigned tasks. Yeah, sometimes you miss something, and another field leader can help you out, but that should never be your normal mode of operation. I've worked with Mike for years, and he puts in a lot of time making sure he has a solid plan and that his crew has exactly what they need when they start a new task. You should be paying attention to how Mike does things. He's been a successful field leader for many years. Now, I suggest you get back up to the third floor and have your crew start taking apart that pipe rack before the plumber decides to back-charge us."

11:08 am

On the ninth floor, Gary's journeyman and apprentice are attempting to install conduit in the slab before the concrete pour the next morning.

Journeyman: "Hey, when you were working on the other decks, did you guys have a harness and retractable lanyard?"

Apprentice: "Yeah, we did. But I haven't seen one since then, and that was a few weeks ago."

Journeyman: "Well, we need to get this conduit stubbed up out there by the edge of the building. You stay over here. I'll just run over there really quick and get it done."

Apprentice: "I'm not sure that's a good idea."

Journeyman: "You heard what Gary said about being in a rush to get this installed before the pour tomorrow morning. We don't have time to search the entire job site for a harness and lanyard. Plus, I need this job. I don't want Gary to think I can't get something done when we're up against a deadline."

Case Study ...continued...

11:32 am Gary gets a call from the general foreman requesting that he immediately come to the job trailer. When Gary enters the trailer, he sees the general foreman along with both the general contractors' superintendent and the safety representative. Also standing nearby are his journeyman and apprentice whom he sent to the ninth floor.

General Foreman:	"Would you like to explain to everyone why you would send your crew members to work on the deck without making sure they had all of the proper safety equipment to do their job?"
Gary:	"What? What are you talking about?"
General Foreman:	"These two were caught up on the ninth floor working on the deck without a safety harness."
Gary:	"They were?! That's not my fault! I asked them to work on the deck this morning, but I never told them to do it without a safety harness!"
General Foreman:	"You didn't answer my question. Did you make sure they had all of the safety equipment they needed readily available? Did you communicate clearly what your expectations were regarding doing the work safely?"
Gary:	"How am I supposed to be on the third floor and the ninth floor at the same time making sure everyone is doing each task the safest way possible?!"
General Foreman:	"You've put yourself and these two crew members in a very difficult position."
Superintendent:	"You know from our safety orientation that we have a zero-tolerance policy when it comes to flagrant safety violations such as this. We've asked your company to remove both you and your journeyman from this project immediately. Your apprentice will only be written up because he wasn't directly involved, other than watching what the journeyman was doing from a distance."
General Foreman:	"Round up all of your things and leave the project."

TAKE A CLOSER LOOK

1. How did Gary's poor project coordination directly lead to a loss of efficiency and most likely impact the project financially?
2. Did Mike and the general foreman treat Gary unfairly in regard to sharing tools?
3. How does this case study highlight the importance of planning ahead?
4. How does this case study highlight the importance of project coordination?
5. How does this case study highlight the importance of safety?

Case Study ...continued...

How Can an Effective Field Leader Coordinate Projects?

If a crew needs more direction regarding site safety:

- Communicate clearly and often that safety takes priority over production.
- Make it a priority to have safety equipment readily available that is in working order.
- Set an example by always wearing all personal protective equipment (PPE).
- Discipline crew members who do not follow safe work practices.

If appropriate documentation is kept organized:

- Necessary changes to a trade's installation schedule can be made to accommodate another trade's installation based on RFIs.
- A change order can be written to recoup the costs if there is a financial impact to a trade.
- Mistakes and rework can be minimized.

If production challenges occur:

- Efficiency is increased when crew members of each trade can find what they need quickly.
- Keep a thorough record of loose ends and assign crew members to tie them up.
- Give the crew access to the prints so they can double-check details and eliminate preventable mistakes.

PERSONAL PERSPECTIVE

1. Describe a situation in which a field leader made safety a priority on a job site.

Case Study ...continued

PERSONAL PERSPECTIVE

2. Describe the role that documentation played in helping two or more trades coordinate installations on a job site.

3. Role-play how a field leader dealt with a production challenge on a job site.

4. Describe the role that stress played in a journeyman's decision to not become a field leader.

chapter 5

COMMUNICATION AND MORALE

A field leader may be able get the job done without being an effective communicator, but full leadership potential is reached only when good communication skills are practiced. People like working for and with effective communicators. To complete a job on time and within budget, a field leader effectively communicates with everyone up and down the project hierarchy.

An effective field leader clearly communicates tasks to crew members to maximize productivity, is supportive of the crew, and listens to their feedback to improve efficiency. An effective field leader also communicates with other trade field leaders and addresses disruptive behaviors and conflicts in order to keep morale on a positive track.

Section 5.1: Effective Communication

- Identify the five communication skills used to send and receive messages.
- Communicate tasks clearly using effective speaking and listening skills.
- Communicate clearly using written documents.
- Demonstrate effective communication skills with crew members and field leaders from other trades.

Section 5.2: Crew Support and Morale

- Describe eight ways to offer crew support that can improve morale.
- Describe five morale busters that directly impact project success.
- Explain the importance of maintaining crew morale.

Section 5.3: Disruptive Behaviors and Conflict

- Describe four common types of inappropriate conduct.
- Address disruptive behaviors and conflicts caused by differing points of view, difficult workers, and other trades.
- Explain how different tactics can be used to address conflict.

Learner Resources
ATPeResources.com/QuickLinks
Access Code: **335421**

Section 5.1 EFFECTIVE COMMUNICATION

A primary function of an effective field leader is to communicate all the knowledge compiled about a project to the crew. An effective field leader explains each task thoroughly. Relying on vague dialogue, hand gestures, or arm waves to communicate information are poor substitutes for good layout. A field leader who says things like, "Just do it the way we did it that one time" is begging for trouble. Crew members can walk away thinking they know exactly what and where, and the field leader can walk away fully unaware that the crew may install something incorrectly or in the wrong location. It only takes a few missed details when sharing information to cause mistakes or rework. For the sake of efficiency and production, an effective field leader cannot allow this to happen.

A field leader needs to be an effective communicator to positively interact with the crew, other trade field leaders, supervisors, and owners/customers. A field leader must clearly understand the expectations of the project manager and owner/customer. A field leader must also read and interpret many types of documents. A field leader must clearly communicate expected results.

A field leader needs to be an effective communicator to positively interact with the crew, other trade field leaders, supervisors, and owners/customers.

It is also important that the field leader understand information from the crew about what is needed and any problems they foresee with an installation. The field leader's ability to understand the needs of the crew is just as vital as giving them the information they need to do their job. It is also important for the field leader to make it clear to the crew that effective communication is vital to the success of the project. The crew needs to know that no matter how insignificant a question or concern may seem, it is always worth the field leader's time to clarify anything that is unclear.

Every form of communication involves both a sender and a receiver of a message. The sender may speak or write the message. The receiver may listen to or read the message. A message is received when the receiver provides feedback that the message has been clearly understood. Five communication skills are used to send and receive messages: speaking skills, listening skills, observation skills, reading skills, and writing skills. **See Figure 5-1.**

Speaking Skills

Effective speaking skills are used to share information, give instructions, and provide feedback. Speaking skills include using the right words to convey a message and pronouncing each word clearly and at an adequate volume so that the chance of a misunderstanding is minimized. For example, giving the direction, "I need you to mount that one panel on the fourth floor" is less clear than, "In the east electrical room on the fourth floor, I need you to mount panel E412C at 78 inches to the top of the can."

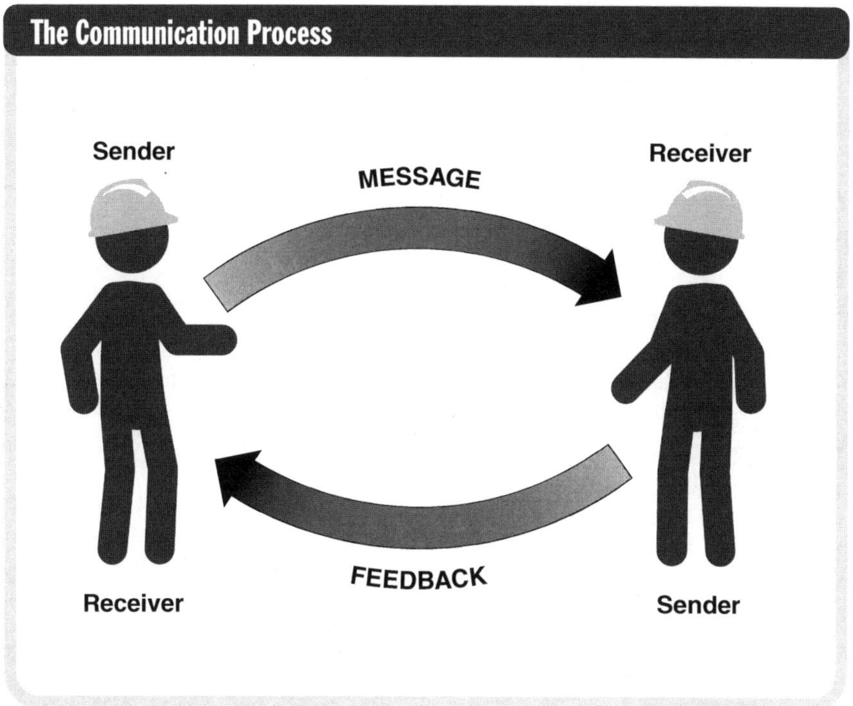

Figure 5-1. Communication skills are used to give and receive instructions, share information, and provide feedback.

Leading in the Field: Do It Like That

The field leader of a framing crew could say, "Go over there and frame in that column. Frame it like that one Joe framed on the third floor," and then point in the general direction of the task. Or the field leader could describe the task in the simplest, most precise way and get the desired results. For example, "I want you to frame column 5B. If you have any questions, ask Joe. He framed one just like it on the third floor."

Though it may be easy for the field leader to visualize column 5B and the specifications detailing how it needs to be framed, the crew members might not have the same vision. It does not cost more time or energy to clearly communicate and avoid potential confusion.

A field leader can improve speaking skills by always using the appropriate tone and volume level. If a field leader shouts orders to the crew in a harsh tone, the field leader is not demonstrating effective speaking skills. The use of slang and foul language is very inappropriate. An effective field leader clearly explains tasks in stages and uses prints, sketches, and related documents when appropriate.

Giving Clear Direction

Explain Tasks Clearly. Throughout the planning process, the field leader commits many details to memory. An effective field leader remains mindful of this fact when sharing information. If not, important details may be glossed over in attempts to keep conversations short. Cutting explanations short leaves it up to the crew members to interpret the details of the job based on guesswork instead of receiving answers to their questions. Rushing when sharing information is just poor communication and guarantees wasted labor hours and loss of productivity.

Explain Tasks in Stages. An effective field leader explains a task in stages, stopping at different points along the way to make sure the information is understood and periodically asking the crew if they have any questions. **See Figure 5-2.** This enables crew members to paint a vivid picture in their minds of the steps they need to follow. It also allows crew members to ask questions that may identify something that has been overlooked, which in turn provides the field leader an opportunity to address the oversight before work begins.

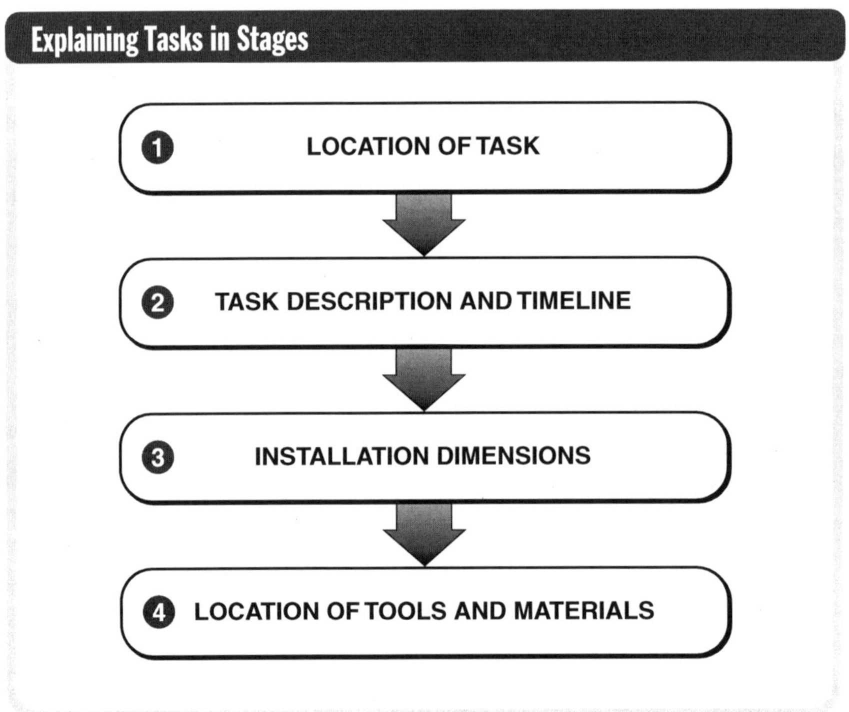

Figure 5-2. An effective field leader explains a task in stages, stopping at different points along the way to make sure the information is understood and periodically asking the crew if they have any questions.

Use Documents to Explain Tasks. Communicating in a manner that is best suited to the listener is the attention to detail that makes an effective field leader. Everyone does not learn the same way—some people are more visual, some are more auditory, and some are more hands-on. Adding a print or a sketch to back up verbal directions increases the likelihood that an installation will be done according to specifications. This additional form of communication increases accuracy and helps to eliminate unnecessary mistakes, a combination that leads to a positive production rate for a project.

Listening Skills

Listening

Listening skills always seem to be in short supply. Often times, a conversation will be interrupted by a speaker asking, "Are you even listening?" This is usually because the intended listener was "hearing" but not "listening" to what the speaker was saying. Effective listening skills require active listening to confirm the message is being received. An active listener makes eye contact with the speaker to indicate that the message being conveyed is valued and also provides feedback to the speaker using gestures such as nodding the head to show understanding. An active listener shows respect by not interrupting the speaker. Active listeners also summarize what they hear and confirm it with the speaker to reinforce the message and allow for any needed clarification.

When a field leader communicates with a crew member regarding a new task, it is important to listen carefully to what is said and to provide thorough and thoughtful feedback or answer any questions. Likewise, discussions regarding an installation should never be a monologue. A true collaboration of thoughts and ideas can only be accomplished when both parties carefully listen to each other and come to agreeable conclusions.

Every tradesperson that a field leader works with holds a unique wealth of knowledge and a perspective that should never be taken for granted. People who feel engaged and who receive the respect they deserve are much more likely to share their wisdom. Active listening is a sure way to inspire crew members to help run a project efficiently.

Leading in the Field: Rushed Communication

A field leader has to get to a coordination meeting, but also needs to explain a task to crew members so that the task can be started during the meeting. In a rush, the field leader attempts to explain the task in abbreviated phrases. "You need to get some pipe in the air right away. The flooring contractor is going to be in here in a couple days, so we will not have access to this space for a while. The priority is to get the branch conduits stubbed out of this room. You know what to do. I have to run."

As the crew members listen to the disjointed directions, they become less and less engaged because they are not receiving the details needed to make a proper installation. Confusion mixed with disinterest sets in. One crew member tries to interrupt the field leader to ask a question, but is cut off. Multiple questions form in their minds as the field leader hurls bits and pieces of information at them, never leaving an opening for them to interject. They slowly lose eye contact with the field leader and begin looking at each other in hopes that someone on the crew understands.

Upon returning to the task location the next morning, the field leader finds that the conduit sizes installed are incorrect, the panel that the conduits are coming from is the wrong one, and the conduits are leaving the space at the wrong elevation and are heading out of the wrong side of the room. In the end, it is the field leader who is responsible for the labor hours lost due to the rework that is now required. A field leader with effective communication skills greatly reduces the risk of making this type of avoidable and costly mistake.

Welcome Questions. A field leader who uses effective communication skills encourages others to ask questions while listening to information about a task. This clearly indicates that the person is interested in knowing as much about the task as possible in order to minimize mistakes, which means less wasted labor. An effective field leader will also try to anticipate any question or detail needed for the installation so that questions can be answered as they come up.

Crew members do not want anyone to think that they do not know what they are doing. Crew members want their field leader to have faith in them. An effective field leader avoids making anyone feel uncomfortable for asking questions. Doing so only discourages further questions and costs the project time and money in the long run.

Provide Answers. If the answer to a crew member's question is not known, the field leader needs to take the time to investigate and get back to the person with a correct answer. Leaving any question unanswered is unacceptable. All it takes is one unclear statement for the rest of a conversation to be lost. An effective field leader would say, "That is a good question. Let me do a little more research and get back to you."

A field leader who views questions as a threat rather than as an attempt to learn information looks foolish. An effective field leader is not embarrassed by questions that cannot be answered immediately. And making up answers is never a solution. A field leader cannot possibly memorize everything there is to know about a project. So it is better to admit not knowing the answer than to have to apologize for giving out incorrect information.

Observation Skills

Observation skills involve interpreting the nonverbal cues of body language. A great deal of communication is conveyed through nonverbal cues, such as posture, facial expressions, and gestures. **See Figure 5-3.** When communicating, an effective leader looks for nonverbal cues that usually indicate whether or not someone is present in both mind and body or just in body. A person who does not make eye contact, seems distracted by things, or seems overwhelmed is saying something without uttering a word.

A nonverbal cue that is consistent with a verbal message reinforces meaning. However, correctly interpreting nonverbal cues can be a challenge. For example, an effective field leader knows that head nodding is not necessarily an indicator of understanding. To confirm the message, the field leader should ask the crew member to repeat the task to confirm correct understanding before ending the conversation. Observing nonverbal cues can also provide useful information to aid communication. For example, if a field leader observes expressions or gestures of frustration from a crew member, speaking to that person with encouragement rather than criticism will yield better results.

When a field leader is explaining a task to a crew member and the crew member begins to look away as if losing interest in the details, the crew member may be confused or concerned about something. As soon as a field leader observes that the crew member's attention has been lost, it is important to re-engage the crew member before going further.

For instance, an effective field leader might stop and say something like, "Asking questions is the best way to get the job done right." An effective field leader may also ask the crew member to restate the task to check for understanding.

Nonverbal Cues

Figure 5-3. A great deal of communication is conveyed through nonverbal cues, such as posture, facial expressions, and gestures.

It is important to remember that most people are not comfortable admitting they are having trouble understanding what they are being asked to do. They often need to be told that is okay to ask questions. A field leader admitting to a crew member that it is possible that the instructions given were not clear can also be valuable. Showing a willingness to own part of the responsibility for miscommunication puts listeners at ease. It may even encourage them to ask more questions and result in better productivity due to clear direction.

Reading Skills

Reading skills enable a person to obtain information that often contains more detail than can easily be expressed by speaking. From prints to technical manuals and purchase orders, there are many forms of written communication that a field leader reads in order to receive valuable information. A field leader can spend hours, days, or even weeks putting together a detailed plan for a project. The field leader is constantly reading and deciphering prints in depth, while also gathering dimensions, selecting and ordering materials, and communicating with other trades. This working knowledge of a project's many facets is something the crew, as a whole, will never possess. There are not enough labor hours in any project to have the entire crew spend large amounts of time reading prints to notice every detail. In most cases, crew members will spend only 5% to 10% of the time that a field leader spends reading and studying prints.

Leading in the Field: Reading Prints and Specifications Carefully

A field leader is in a rush, unprepared, and needs to find something to keep a couple of crew members busy. The field leader quickly scans through the prints to find a task for them to do and then heads out to the job site to lay them out.

In the midst of the field leader's explanation, one of the crew members suggests an idea regarding the installation. The field leader listens. The crew member provides a similar yet different idea that will use a different type of material but will cut down on the labor hours needed to complete the task. The field leader thinks about it for a minute and decides there is no reason not to do it, and tells them to get started right away.

A couple of days later, the general foreman contacts the field leader to question the installation. The general foreman reinforces that, on the prints and in the construction specifications, there are very strict guidelines for the approved materials and methods of installation, and that the installation needs to be redone because the guidelines were overlooked. The field leader has to ask the crew to remove the installation and do the task over. Multiple labor hours are wasted in the process.

Both on the surface and buried deep within construction prints and specifications are very precise details that cannot be disregarded. An effective field leader takes the time to read all of the documents carefully and then details each installation thoroughly before directing the crew.

If the information being read is unclear, the reader must request clarification or seek answers using relevant documents. For example, after reading a print, a field leader might complete a request for information (RFI) and send it to the architect for clarification, or may search online for more information about an unfamiliar symbol on a print.

Writing Skills

Whether a written message is delivered via a print, a form, a technical manual, a text, or an email, clarity and context are necessary for the reader to understand the message. The reader of a written communication does not have the benefit of hearing the tone of the speaker's voice or seeing facial expressions, so it is very important to write clearly in order to accurately convey information and avoid misunderstandings. To do this, the writer must determine the purpose of the message and then use appropriate words, proper spelling, and grammar. It is also important to write legibly so that the message can actually be read. Daily job journals, change orders, and RFIs are examples of written communication used on construction jobs.

When a field leader proofreads a written document for clarity and context, it helps to ask the following questions:

- Are the relevant print/specification documents mentioned?
- Will the reader come away with a clear understanding of what is being asked or explained?
- Are there any missing details that will initiate follow-up questions to be asked before the original question can be answered?

For example, an RFI should indicate the items that need to be clarified for the job to move forward. An unclear RFI will only result in further confusion and loss of production time.

Leading in the Field: Written Inquiries

A field leader finds a discrepancy on a print and needs to get clarification from the architect before moving forward on a designated task. An RFI needs to be written, but time constraints and distractions make it difficult. The field leader sets the RFI aside and writes down a few details and questions to review later while proofreading the document. Then later that day, the field leader remembers that the RFI was never sent, quickly decides it is good enough, and sends it.

A couple of days later, shortly before the crew expects to begin the designated task, the field leader receives a response from the architect. Unfortunately, the architect admits to not being able to read some of the scribbled notes, and the legible ones were short on details, which only prompted more questions. Plans for moving forward with the designated task now have to be postponed and the RFI has to be rewritten and resubmitted to the architect. Rushing an unclear written communication always costs time and money in the end.

The crew counts on the field leader for accurate information about installations. Supervisors count on field leaders to run the most efficient and productive job possible. Owners/customers need to believe the field leader is fully qualified to manage and execute the job. Vague written communications make a field leader appear unprofessional or uninformed or both, and lessen the field leader's credibility with crew, supervisors, other trade field leaders, and owners/customers.

Daily Communication

A field leader uses speaking, listening, observation, reading, and writing skills to effectively communicate with many people on the job. **See Figure 5-4.** An effective field leader is constantly assessing the quality and content of all forms of communication as they relate to installations on a project, whether explaining a task to a crew member or coordinating with another trade. An effective field leader makes a concentrated effort to model effective communication skills and establish their value to the success of the entire team on every project. Each day, an effective field leader checks in with the crew and cooperates with other trade field leaders to get the job done.

Effective Daily Communications

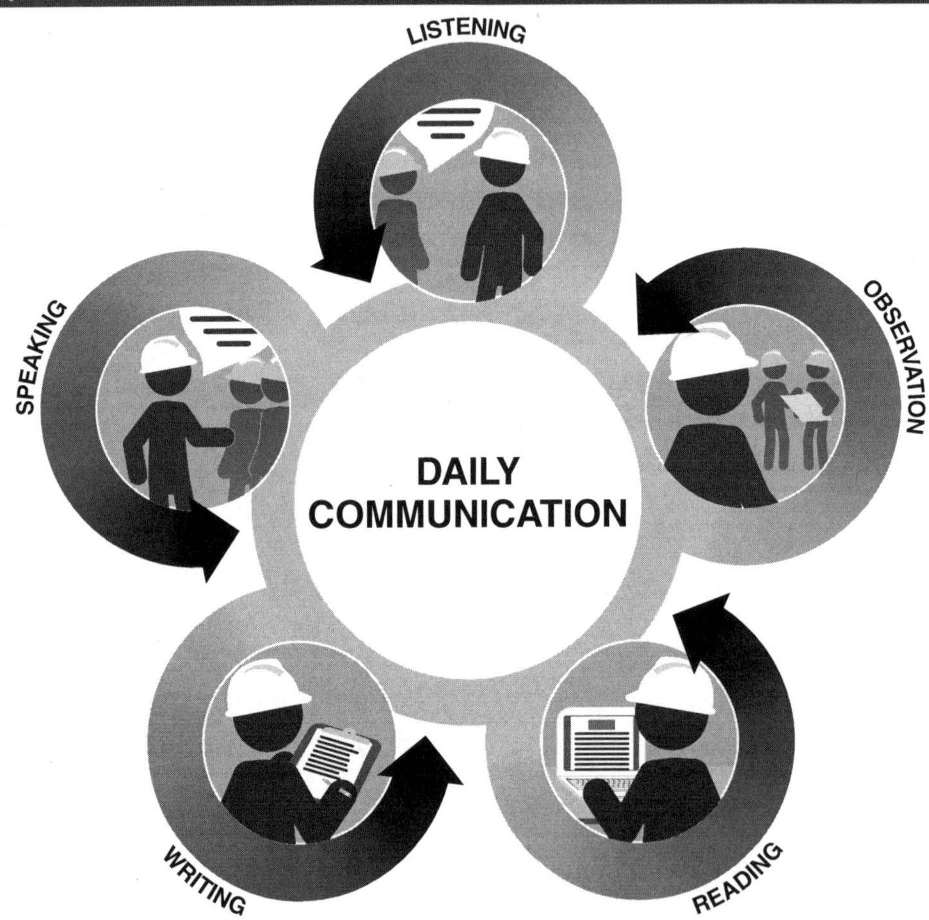

Figure 5-4. A field leader uses speaking, listening, observation, reading, and writing skills to effectively communicate with many people on the job.

Check in Often with Crew Members. New questions are sure to arise throughout the installation process, and checking in with crew members gives the field leader a head start on finding answers that may not be immediately available. An effective field leader should provide crew members with every opportunity to ask questions that inevitably will arise. A field leader enters the space where the crew is working, asks if everything is going well and if there are any questions, and then moves on to the next task. But remember, it is important that the crew not perceive the field leader as distrusting of their abilities or work ethic.

When a crew member is ready to move to a new task, an effective field leader will find time a few days prior to give the crew member information. A few advance details will allow the crew member to start thinking about materials that may be needed or information that may need clarification. Doing this also facilitates the task being done efficiently. Although there never seems to be enough hours in the day, taking the time to explain new tasks in advance promotes effective communication and leads to project success.

Coordinate with Other Field Leaders. The importance of effectively communicating with other trade field leaders cannot be overstated because it is critical to every field leader's success. Projects end up in disarray when every trade is not on the same page on everything from scheduling to installations, which requires establishing a common direction and realistic completion goals. It is also important for the field leader to be flexible and to make an honest effort to let the other trades know if a problem exists so that they will have enough time to adjust their plans and schedules.

Think about what would happen if, instead of communicating, all the trades came rushing into a project with only their respective installations in mind. When a field leader does not think about how a specified installation may affect another trade, materials pile up, the site becomes disorganized, tensions run high, and schedules get missed. **See Figure 5-5.** To avoid this scenario, each trade field leader must use effective communication skills to keep the other field leaders informed. When one trade falls behind schedule, it impacts all of the other trades on the job.

The Domino Effect

Figure 5-5. When a field leader does not think about how a specified installation may affect another trade, materials pile up, the site becomes disorganized, tensions run high, and schedules get missed.

Most field leaders are willing to work with other trade field leaders who are up front and honest. However, if a field leader has a habit of waiting until the last minute, the other trade field leaders will become frustrated and be less inclined to adjust their plans. Likewise, walking around with blinders on and ignoring the needs of other trades makes it virtually impossible to complete projects on time and within budget. Field leaders who believe their installations should take precedence over those of other trades are shortsighted.

If one bad seed is present on a project, effective field leaders who value communication can band together to prevent a tension-filled environment from forming and affecting the success of the project. Communication is essential, and only a group effort will make the job a success.

Section 5.1 For Discussion

1. Explain how clear directions are communicated.
2. How does active listening aid communication?
3. Explain the importance of observing nonverbal cues when communicating.
4. What role do reading skills play in a field leader's success?
5. Explain the value of clearly communicating on written documents.

Section 5.1 Take Action

1. Role-play a field leader clearly explaining a task to a crew member.
2. Role-play a field leader and a crew member being active listeners.
3. Complete an RFI and ask a partner to read it for clarity.
4. Role-play two field leaders from different trades communicating at a coordination meeting.

Section 5.2

CREW SUPPORT AND MORALE

A construction field leader interacts with individuals of many different backgrounds, skill levels, mindsets, and personalities. Finding a way to work well with a variety of different individuals while encouraging them to reach their potential builds respect and admiration and, in turn, increased productivity. A field leader's success is directly tied to the success of the crew, whose success is based on having everything they need from tools and equipment to support and motivation. Being a field leader is about setting a crew up to meet or exceed the goals set out for them.

> A field leader's success is directly tied to the success of the crew, whose success is based on having everything they need from tools and equipment to support and motivation.

There are field leaders who possess every technical skill necessary to build just about anything but who remain stagnant in their careers. Conversely, there are field leaders who are less talented with the technical aspects of the job but are promoted. A majority of contractors understand that the most technically sound craftsmen do not automatically make effective field leaders. They know that it takes strong technical skills coupled with the ability to inspire a crew to work hard.

Effective field leaders know that disrespecting crew members or ruling them with an iron fist and still expecting them to perform at peak performance never works. Treating people poorly lowers morale and discourages them from caring about project goals. In fact, some people may even go out of their way to see that the field leader does not reach these goals.

Crew support and morale should never be undervalued due to their effect on the success of a project. Both profits and success in the construction industry are directly related to productivity. So it only stands to reason that if the people who have the power to increase or decrease production are engaged and motivated, a field leader's chance of running a successful project remains intact. There is also no doubt that keeping the crew involved in the process also pays long-term dividends.

Boosting Morale

No one wants a job to be an unbearable grind. Being part of a crew that works for a field leader who is known to tackle the toughest projects successfully inspires motivation, pride, and ownership. People who work hard, get along well, and feel like part of a team will find their jobs gratifying.

There is no perfect recipe to boosting morale. Project variables and changing crew dynamics make it impossible to use what works from one project to the next. With that said, the tried-and-true way to empower a crew to do a great job and get desired results starts with mutual respect and ends with good communication.

Motivating People

Following are some simple rules for motivating people on the job site. Applying these simple solutions will not cost a company any money and does not require unreasonable amounts of time. The best part is that a committed and hardworking crew will be the reward.

Trust People. People who feel trusted feel empowered. An effective field leader gives individuals the benefit of the doubt until they prove themselves to be untrustworthy.

Compliment People. Nothing goes further to motivate individuals than a genuine compliment from their field leader. People are proud of the work they do and feel appreciated when their efforts are acknowledged. It is important to be sincere. Lavishing unwarranted praise only spawns resentment.

Give Credit. Always give credit where it is due. Most construction professionals spend many years learning their crafts. An effective field leader never takes credit for something a crew member does, especially when a supervisor, another trade, or a customer is present.

Show Appreciation. Saying "thank you" or "great job" can keep an individual engaged and motivated. The outcome is well worth the conscious effort involved.

Share Information. All it takes is a couple of minutes every few days to let the crew know how well things are working according to plan and where the job is headed. A crew wants to feel they have job security, and their field leader is the only one who can give it to them.

Leading in the Field: High-Quality Installation Request

The customer for a project demands a high-quality installation and wants to have a building that is a source of pride for years to come. How can a field leader motivate the crew to make a high-quality commitment?

He could approach the crew before they go to work or right after their lunch break and say, "First of all, you need to know that you're doing a great job and the customer is happy. The work looks great, and I appreciate everyone's efforts. Now that some of you are gearing up to start running exposed conduit, I'm asking you to continue to make it a professional installation. This is an important customer to our company who expects the best from us. Let's continue to show the customer what we're capable of. Keep up the good work!"

A few words of praise only take a minute, and there is little chance the crew will complain. Addressing the entire crew in a positive manner is a good way to get the results desired.

Inspire Teamwork. Crew members have different expectations, interests, needs, and ability levels. But a field leader can only be effective when the crew works as a team. By setting a good example and treating each person equally, a field leader fosters mutual respect and collaboration and shows how everyone can work together toward a common goal. But if a field leader shows favoritism toward one or more crew members, the others may take notice. Those who are not favored will quickly become discouraged.

Offer Feedback. Many field leaders are uncomfortable providing constructive feedback to crew members who are not pulling their weight. It is never easy to tell someone they are not doing a good job. But if a field leader lies about, embellishes, omits, or avoids the truth while talking to a crew member, the crew member will know the field leader is blowing smoke. Not only is it insulting, but the field leader will lose the crew member's respect.

Leading in the Field: Disappointing Results

Two members of a crew of 15 are consistently failing to get tasks done. The field leader asks these two individuals if there is anything that can be done that will increase their productivity. Unfortunately, no progress is made, and they continue to fall short of expectations.

Some field leaders may decide to spend a little more time checking up on them, but such an approach usually backfires. Taking on the role of a snoop makes other crew members think the field leader is trying to discover everyone's secrets instead of just focusing on the two who are not pulling their weight. If there is one thing that shuts a crew down, it is a field leader who assumes that if a couple of people are slacking off, then the rest of the crew must be slacking off too. The entire crew will feel threatened and resent the field leader even more than the people who caused the problem.

Most people are honest, hardworking, and willing to put in a day's work. However, a crew member with a bad attitude can poison the morale of an entire crew. Deal with the individual as soon as a problem is detected. Be very direct and make it clear that if the crew member cannot perform to the expected level, there will be no job to come to. This is a fair approach that gives the crew member the opportunity to make the requested adjustments. Most people will respond by working to improve their productivity.

When a crew member needs to work on productivity, an effective field leader will offer some ideas for improvement and then offer the crew member support to achieve those goals. The longer a field leader works with an individual crew member, the more opportunities the crew member has to build both knowledge and skill.

Encouraging initiative helps each individual take personal pride and ownership in a project. And in turn, this environment of responsibility results in a crew who pushes themselves to excel because they want to, not out of the fear of being fired.

And remember, motivation is never accomplished through intimidation, threats, or condescending dialogue. Threats and intimidation do not inspire loyalty or a desire to complete a project on time and within budget.

Leading in the Field: My Way or the Highway

The field leader of a project notices that a prompt crew member with a good work ethic is making mistakes, taking longer than usual to complete tasks, and appears to be out of sorts. Because the field leader has a "my way or the highway" philosophy, the crew member is told to kick into high gear or get packing. A week later, the crew member is gone.

How was this approach flawed? A field leader cannot always be in the same location where the crew is working. That means there is no way of knowing why production might be lacking. A field leader can make a snap judgment based on assumptions without really knowing the facts. This approach is wrong more times than not.

A field leader is not a motivator when assuming that the only reason a crew member is not completing a task on schedule is because that person is slacking off. There may be legitimate reasons for productivity being down. For example, the worker's poor production may be because the work area is congested with workers from other trades. The responsibility lies squarely on the field leader's shoulders to make sure there are no obstacles hampering the crew's performance.

A rushed judgment can also cause collateral damage. When the rest of the crew finds out how the situation was handled, they could view the field leader as someone who jumps to conclusions. This is no way to build respect, trust, and loyalty from a crew.

Offer Recognition. An effective field leader knows that each project requires a different form of motivation, not only because the landscape and building design have changed but so have the members of the crew. What works to motivate a crew on one project may fail with another crew on the next project. This requires the field leader to constantly shift gears because people are motivated in different ways. There are many different ways to recognize people.

Addressing the crew with a "hooray speech" every couple of months to pump them up is usually not effective. Likewise, buying a stack of pizzas for lunch will not be wasted on the crew, but their appreciation only lasts until lunch is over. Handing out company promotional items such as t-shirts, hats, and coffee mugs is appreciated. But owning a coffee mug or t-shirt does not mean that the crew member will turn around and go back to work with renewed passion for company success. Handouts do not build loyalty the way words and actions can.

Some companies offer prizes based on the completion of a phase of a project, on production and performance, or on the accomplishment of a set amount of labor hours without any time lost due to injury. While it is always nice to receive appreciation for a job well done, this method may lead to under reporting of incidents and usually fails to motivate any change in work habits. Motivation requires ongoing work, not just a carrot at the end of a stick.

The best person to motivate the crew is the field leader. An effective field leader recognizes individuals who carry more than their fair share of the weight on a project. The field leader will find that the motivational level of the entire team rises when connecting directly with individual crew members.

Morale Busters

A crew's perception of a field leader's commitment to their success is what forms their morale. Working for a field leader who has great technical skills but lacks leadership skills can be frustrating and discouraging, because a crew looks to their field leader for direction and assurance. If a field leader remains positive and upbeat even when things are not going as planned, the crew will go along for the ride.

Think about the conversations the crew is having while they are working, on break, eating lunch, or having a beer after work. All it takes is for two or more crew members to show up at a gang box at the same time looking for tools or materials that are not there for the first sign of trouble to appear. A couple of people commiserating about how ironic it is that the field leader has high production expectations but does not give them what they need is all it takes to get poor morale rolling. Poor morale starts with irritation that becomes frustration and can eventually snowball into animosity and infighting. **See Figure 5-6.** And as the snowball rolls downhill and gets bigger and bigger, more people get drawn into the conversation. Soon the field leader has been branded as someone who does not care, and it will take a lot of effort to prove otherwise.

The Snowball Effect

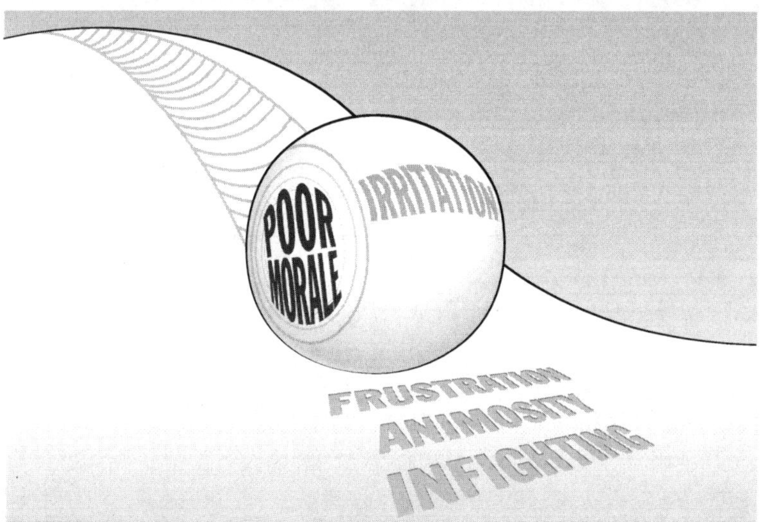

Figure 5-6. Poor morale starts with irritation that becomes frustration and can eventually snowball into animosity and infighting.

Effective field leaders are able to maintain crew morale when difficulties arise. Following are common situations that often lead to crew dissatisfaction and low morale, along with tips for how to avoid these pitfalls.

Alienation. Without the recognition they deserve and need, crew members will feel alienated. And if the crew is not getting their due in terms of recognition, they may assume that the field leader is taking credit for their hard work. Some field leaders take the position that if they are not receiving credit from their supervisors, they do not need to recognize the people who work directly for them. This attitude is counterproductive and serves no one.

A field leader is responsible for making sure the crew gets the credit they deserve when they do a good job. If a field leader takes the praise the crew deserves, the crew will not feel appreciated. Effective field leaders put aside their own need for recognition and focus on the needs of the crew.

Missing Tools and Materials. The majority of people show up every day with the intent of putting in a solid day's work. However, if the field leader does not keep the job stocked with tools and materials, the crew's opportunity to be efficient and productive is lost. This type of work environment causes them to lose interest in the project. From their perspective, it appears that the field leader does not care if they are efficient. Actions always speak louder than words. No matter what a field leader says, it is the example that is set that matters.

> If the field leader does not keep the job stocked with tools and materials, then the crew's opportunity to be efficient and productive is lost.

Borrowing from Other Trades. Although there will be times when borrowing materials or equipment from another trade may be necessary, this approach is typically used by field leaders who do not plan. Not having what is needed to do a job grinds down a crew's morale and causes them to resent the field leader. Some crew members may even go so far as to hide tools or materials because they want to be able to find them when they need them again.

Every time a specialty tool, a ladder, a lift, or any other piece of equipment is borrowed from another trade, the field leader has to hope that the other trade will not need it while the crew is using it. Consider how this type of poor planning can affect the profitability of a project. First, there is lost time as the crew member looks for a tool but cannot find it. Second, there is lost time as the crew member finds the field leader and reports that the needed tool cannot be found. Third, there is lost time as the crew member searches for another trade who might have the needed tool.

Suppose the other trade leader takes back the borrowed item early. Now the crew member has to stop again and start the process of finding another tool to borrow. Even more labor hours are lost. Keep in mind that this example only takes into consideration the lost time of one crew member and one event.

To get a true picture of the costs, multiply the results by the number of crew members who borrow from other trades due to their field leader's lack of planning. Depending on the size of the crew, the loss due to unproductive downtime could be anywhere from an hour per day to many hours per day. **See Figure 5-7.**

Job-Site Politics. Politics exist on every job site. However, job-site politics should not involve the crew. For example, if a field leader has a disagreement with another trade field leader and then tells the crew about it, they may decide to cause problems with the crew members of the other trade. Getting the crew involved in job-site politics only exacerbates the problem and brings down morale. An effective field leader discusses conflicts with the project manager or the other trade field leader involved and works to resolve things quickly, while keeping the crew focused on the project.

Borrowing Costs Downtime

Figure 5-7. Depending on the size of the crew, the loss due to unproductive downtime could be anywhere from an hour per day to many hours per day.

Manpower Cuts. One of the most difficult tasks for any field leader is determining who will be let go when it comes time to cut back a crew's size during a project. As the end of a job draws near, it is common for morale to slip because each person wonders who will be cut loose. All construction projects have a beginning and an end, but this fact does not alter the anxiety that comes with the prospect of not having a job.

An effective field leader values the job that each individual does and knows that letting a crew member go without warning will be the last thing that person remembers. An effective field leader also knows that when a crew needs to be thinned out, crew members are very aware of this. The people being let go are co-workers, perhaps even friends. They all worked together to build a project day in and day out. The workers that stay behind to finish the project want to see their teammates treated with dignity and respect. They watch the field leader to see how the process will be handled and adjust their respect for the field leader accordingly.

Maintaining Crew Morale

Motivating a crew is a continuous effort throughout every project. Morale naturally ebbs and flows, but it is up to the field leader to monitor the pulse of the crew and make any necessary adjustments. To do this, an effective field leader cannot be egotistical. A field leader with a big ego is perceived as unapproachable. The crew would rather stick together and gripe behind the field leader's back than voice their thoughts.

A field leader's crew is the first line of defense for identifying problems on a project. An effective field leader makes sure the crew knows that they are being counted on and the job cannot be done without them. A crew appreciates a field leader who wants to help them succeed.

Effective field leaders know that workers who feel valuable, involved, and appreciated are much more likely to have high morale and be productive. An effective field leader spends a few minutes every week with the entire crew to discuss any difficulties they are having, welcomes their concerns and opinions, and helps in finding solutions within the team. During daily or weekly communications with project managers, an effective field leader may tout that some members of the crew are doing an exceptional job. This can be said without putting down other crew members. The fact that the field leader states that some individuals should be recognized gives the project manager the initiative to visit these crew members on site and personally express appreciation for their hard work.

One of the highest accolades a field leader can receive is to be recognized as loyal to the crew, to co-workers, and to the company. Loyalty is the act of showing devotion, trustworthiness, and allegiance to another through consistent actions. Loyalty is a character trait that speaks volumes about who a person is both as an individual and as a leader. For example, when a reputation of loyalty precedes a field leader on a job, any working relationships formed with new crew members, field leaders from other trades, or company management will start out on a solid foundation. People will enter into the relationship willingly and with confidence that their expectations will be met because the field leader is held in high regard.

Leading in the Field: Loyalty

The field leader of a very long project has several individuals on the crew who always go above and beyond what is asked. To express appreciation for their dedication, these crew members are kept until the very end of the project. The field leader also communicates appreciation by frequently thanking them for their contributions.

To recognize their loyalty, the field leader talks to the project manager long before the project is complete about the important role these individuals played in the project's success. The field leader also calls other field leaders within the company and lets them know that several very talented people will be looking for another project soon. Consistently showing appreciation to those who do their job well is a win-win situation. Every time the field leader shows loyalty to an individual, the individual becomes more loyal to the field leader.

Contractors cannot run successful companies without skilled, dedicated, and loyal field leaders. They rely on the integrity of every field leader they employ. How does a person go about building a reputation of loyalty? It is quite simple: by treating people with respect. Effective field leaders respect their crew members and remain grounded. They do their job, and they do it well.

Section 5.2 For Discussion

1. Explain how trust, compliments, giving credit, showing appreciation, and sharing information boost morale.
2. Explain how working as a team increases crew morale.
3. Why is it important to be direct and clear when providing constructive feedback?
4. How do alienation, missing tools and materials, borrowing from other trades, job-site politics, and manpower cuts affect morale?
5. Explain the role of loyalty in maintaining crew morale.

Section 5.2 Take Action

1. Role-play a field leader boosting the crew's morale.
2. Create a form of recognition that would show crew support and improve morale.
3. Describe a situation involving a lack of crew support and morale, and ask a partner to provide constructive feedback that would improve the situation.
4. Role-play a morale-busting situation and how it could be addressed.

Section 5.3

DISRUPTIVE BEHAVIORS AND CONFLICT

No two construction projects are alike. Plans, specifications, site conditions, construction methods, the trades involved, and the goals all vary. Unfortunately, no matter how well people are treated, there will always be a small percentage of individuals who cause disruption and conflict on the job site. It is a field leader's responsibility to isolate disruptive behaviors and conflict and then implement viable solutions to keep the project on track.

Ego, pride, jealousy, competition, and performance issues are the primary reasons people exhibit disruptive behaviors. Poor communication, limited resources, and power struggles are often the root cause of disruptive behaviors. **See Figure 5-8.** Poor communication leads to misunderstandings. A lack of information as well as inaccurate information can result in work being done that does not meet expectations. Conflicts often arise over the use of space, materials, or equipment on the job site. Power struggles often result when an individual or one trade wants to take control of a project.

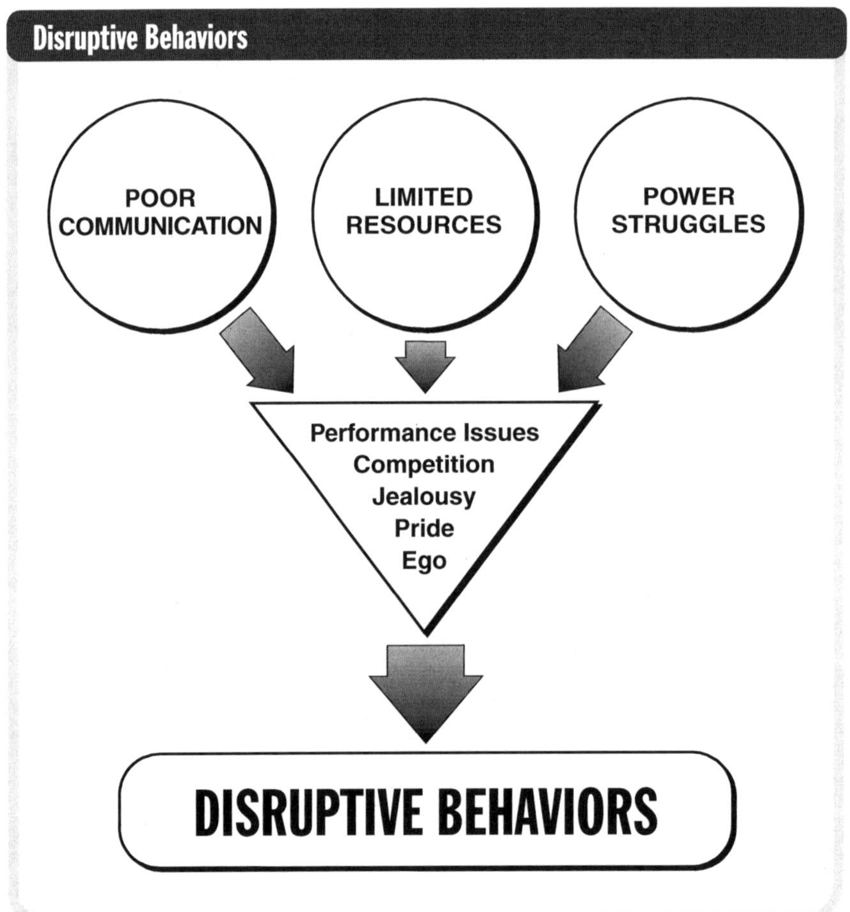

Figure 5-8. Poor communication, limited resources, and power struggles are often the root cause of disruptive behaviors.

Nothing is more frustrating than having a solid plan, a great crew, all the necessary tools and materials to complete every task—and one individual who chooses to be difficult. And it only takes one person to send a project into a tailspin. That is why it is so important for a field leader to deal with disruptive behaviors and conflict swiftly. If the field leader looks the other way, thinking the problem will go away on its own or is uncomfortable with confrontation, this inaction will negatively affect the crew as well as the project.

An effective field leader addresses disruptive behaviors as soon as they come up and works to keep those behaviors from turning into conflicts. Unresolved conflict often results in lost productivity and creates barriers to cooperation. Field leaders who do not deal with conflict eventually see talented crew members walk away in search of healthier job sites.

Conflict Between Crew Members

Inappropriate Conduct

For the safety and fair treatment of everyone on the job site, an employer's policies regarding inappropriate conduct must be enforced by the field leader. The policies may concern topics like tardiness, safety practices, and abuse, but they could also cover other topics. Following are descriptions of four types of inappropriate conduct that can lead to disruptive behaviors and conflicts on the job site.

Discrimination. The unfair treatment of an individual based on characteristics such as gender, race, ethnicity, age, religion, appearance, or disability is called discrimination. Examples of discrimination include treating someone of a different gender or race differently than those who are of the same gender or race. No one wants to be mistreated or viewed as inferior. Field leaders can avoid discrimination by treating all people equally, including customers, crew members, leaders, and managers. Effective field leaders welcome diversity because they appreciate the differences among various individuals.

Harassment. Behavior that is found to be threatening or that is not considered acceptable to the general public is called harassment. Harassment can range from words or actions that make someone feel uncomfortable or demeaned to situations that make someone feel threatened or endangered. Harassment that includes unwelcome verbal or physical conduct of a sexual nature is called sexual harassment. An effective field leader knows that harassment of any kind creates an unfriendly work environment and does not tolerate it.

Hostile Work Environments. A hostile work environment exists when someone feels verbally or physically threatened in the workplace. Violence, physical threats, and intimidation have no place on the job site. An effective field leader addresses these incidents as soon as possible by giving a verbal warning to the guilty party, documenting the warning, and then reporting it to the supervisor. For example, a bully makes others feel unsafe or puts others down to make them feel insignificant. Bullies create a hostile work environment by persecuting others in order to hide their insecurities. This type of intimidating behavior must be addressed before it has a chance to create a hostile work environment.

Substance Abuse. Zero-tolerance drug and alcohol policies prohibit the use of these substances on the job site. An effective field leader knows that the use of drugs or alcohol on the job site creates an unsafe work environment. Once the field leader suspects or knows that someone on the crew is under the influence, it becomes a priority to get it resolved. Any length of time that this issue is ignored creates an opportunity for that person to potentially harm themselves or someone else.

Leading in the Field: Difficult Decisions

A crew member comes to the field leader first thing in the morning and says that one of the other crew members seems to be out of sorts and appears to be a little unsteady. The crew member is worried and does not know what to do.

The field leader immediately goes to where the unsteady crew member is working to make sure the person is all right. During the conversation, the crew member admits to being out way too late with some friends drinking, but now is just tired and will be fine. The field leader has a difficult decision to make. The crew member is told not to work on anything dangerous or go up on a ladder. A warning is also given that from here on out, the crew member needs to show up to work with a clear head or stay home.

An hour later, the field leader gets a call from the general contractor that one of the crew members was cut while using a razor knife to open a box, and the field leader needs to come to the trailer right away. Once there, the field leader finds out it is the unsteady crew member. The general contractor tells the field leader that intoxication is likely the cause of the incident.

In the safety incident review conducted by the general contractor, it is revealed that the field leader was aware that the crew member was not fit for duty. The field leader is given a written warning by the general contractor based on negligence for allowing the crew member to work while intoxicated and put others at risk.

Differing Points of View

A key driver on all construction projects is time. Each trade promises to perform in a way that will meet the time requirements of the project. However, these promises are based on educated guesses as to the abilities of suppliers, vendors, field leaders, and workers to deliver on their promises. Situations and conditions, such as transportation and weather, are also wild-card factors. Changing conditions and diverse expectations set the stage for differing points of view.

Disruptive behaviors on the job site often arise from differing points of view. A simple problem can quickly grow into a disagreement between crew members with differing points of view. If the disagreement is not resolved, it can become a dispute. And unresolved disputes can escalate into conflicts that have to be handled by superiors who do not have direct job-site involvement.

Miscommunication is usually the root cause of differing points of view. If a field leader does not give clear direction, workers are left to decode expectations. Workers who perceive information to be missing often fill in details based on their past experiences. An effective field leader avoids using phrases like "when you get to it" and "whatever you think" and employs clear directives as to who, what, when, where, and why to prevent workers from tripping over each other while doing their jobs.

An effective field leader avoids using phrases like "when you get to it" and "whatever you think."

Difficult Workers

An individual will not change a behavior pattern by receiving negative attention in front of the crew—the person will only become more defensive. An effective field leader schedules a one-on-one meeting away from the crew and gives the individual verbal notice that the meeting will be documented. During the private conversation, the field leader gives the person an opportunity to explain the situation, as the person may have some valid points that should be considered. An effective field leader then offers the person an opportunity to work on changing the unacceptable behavior.

Some workers may have an "us against them" attitude when it comes to the relationship between them and management, which includes the field leader. If that happens, instead of having a problem with one person, the field leader could also have to regain the trust and respect of all the other crew members.

A field leader who enters these difficult conversations with an open mind has an improved chance of success. In contrast, if a field leader begins the conversation with, "If this guy gets out of line, I'll can him on the spot!" nothing will be accomplished. Being aggressive and making threats is not only unprofessional but counterproductive. An effective field leader takes a tough-but-fair standpoint and makes it very clear that there is an issue that requires immediate resolution.

If during the private conversation, the field leader learns that the person is being intentionally disruptive, the field leader may be able to use that information. Because many people expect to hear, "Get with the program or hit the door!" they often are receptive when confronted with the issue head on because it allows them an opportunity to explain their thinking.

After a private conversation with a disruptive worker, an effective field leader documents the date, time, topics discussed, conclusions reached, and each person's responsibility moving forward. **See Figure 5-9.** The field leader also informs the supervisor of the situation. If the person continues to cause problems after the conversation, the field leader may allow the person one last chance to change direction. In most cases, this will earn the person's trust and respect. However, this is also the conversation where a direct warning is given about the viability of the person's job. This conversation concludes with the person knowing that there will be no more warnings. And even with the best of intentions, positive results do not always occur.

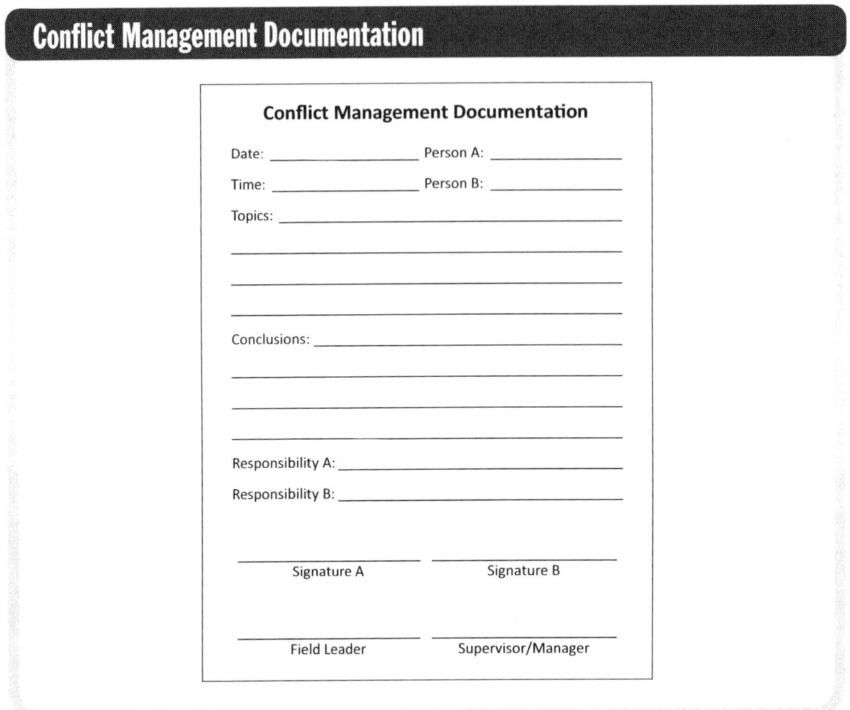

Conflict Management Documentation

Conflict Management Documentation

Date: ______ Person A: ______

Time: ______ Person B: ______

Topics: ______

Conclusions: ______

Responsibility A: ______

Responsibility B: ______

Signature A ______ Signature B ______

Field Leader ______ Supervisor/Manager ______

Figure 5-9. After a private conversation with a disruptive worker, an effective field leader documents the date, time, topics discussed, conclusions reached, and each person's responsibility moving forward.

Verbal and Written Warnings

Addressing Insubordination

Most crew members do their jobs and do them well. Unfortunately, some individuals are unreliable, irresponsible, or even belligerent. An effective field leader uses verbal and written warnings to deter disruptive behaviors.

- Verbal warnings are used to explain the reason that a change in behavior is required. The individual is allowed to explain what happened and then to return to work. A verbal warning allows everyone involved to approach the next task with a positive attitude.
- Written warnings are used to address repeated disruptive behaviors. The seriousness of the situation is explained, and the field leader follows up with a written memo that documents the meeting and summarizes the discussion.

If a crew member fails to change unacceptable behavior after being given both verbal and written warnings, the crew member may be asked to leave. If the individuals directly involved in a job-site conflict cannot resolve a problem at the job site, their supervisors meet to work out a solution. If they, in turn, cannot agree, then the problem is passed on up the management ladder. Immediate terminations may occur when theft, violence, bullying, discrimination, or harassment is involved.

Difficulties with Other Trades

Conflict Between Different Trades

Addressing a difficult field leader from another trade is similar to handling a difficult worker, minus the ability to terminate the person. A productive solution is still based on good communication and getting to the bottom of why a problem exists. An effective field leader starts with a one-on-one conversation the same way as with a crew member, in an area away from other people.

Keep in mind that most people do not see their actions as disruptive. They may even think that they are not the one being difficult. It is important to enter the conversation with an open mind and be willing to listen to the other field leader's perspective. The primary goal is to find common ground to build a working relationship that benefits both trades. Do not be accusatory. State how the issue is affecting progress on the project and describe what could be done differently. It bears repeating that speaking in a respectful way usually yields positive results. Hopefully, the other person will be reasonable and try to accommodate the request.

After the conversation, do not expect things to change immediately. It generally takes just as long to unlearn old habits as it does to create new ones. It may take a reminder of the agreement before change takes place. Be as patient as possible if the person seems to be making a good faith effort.

If after a fair amount of time, however, the person does not cooperate, it may be time to ask for help. Since subcontractors generally work directly for the general contractor, the general contractor has the authority to step in if there is a need for it. If this action does not move things in a positive direction, contact the other field leader's supervisor to see if something can be worked out. The drawback to this step is that any field leader who is reprimanded is going to be hostile to the person who brought up the issue.

Tactics for Addressing Conflict

Whether the people involved in job-site conflicts are field leaders, crew members, or others, they will likely view conflicts differently. **See Figure 5-10.** Some individuals focus on the people involved. Some individuals focus on the results. And some individuals try to balance their focus between people and results. Regardless of viewpoint, a field leader can use the following tactics to address conflict:

- **Accommodation**—The accommodation tactic involves surrendering a position in order to reach an agreeable resolution. Accommodation is useful when the goal of achieving peace is more important than winning the conflict.
- **Collaboration**—The collaboration tactic involves acknowledging everyone's views and taking them into consideration. Collaboration is useful when there is a need to find a resolution to a conflict among multiple individuals or groups.
- **Avoidance**—The avoidance tactic involves delegating controversial decisions to others. Avoidance is a logical choice when it is obvious that someone else is in a better position to deal with the conflict.

- **Contention—**The contention tactic involves winning by operating from a position of power, no matter the consequences. Contention is useful when an immediate decision needs to be made.
- **Compromise—**The compromise tactic involves finding a balance between a desire to win and a concern for the other person's needs and wants. Compromise is useful when people on both sides have equally strong arguments and a project deadline is approaching.

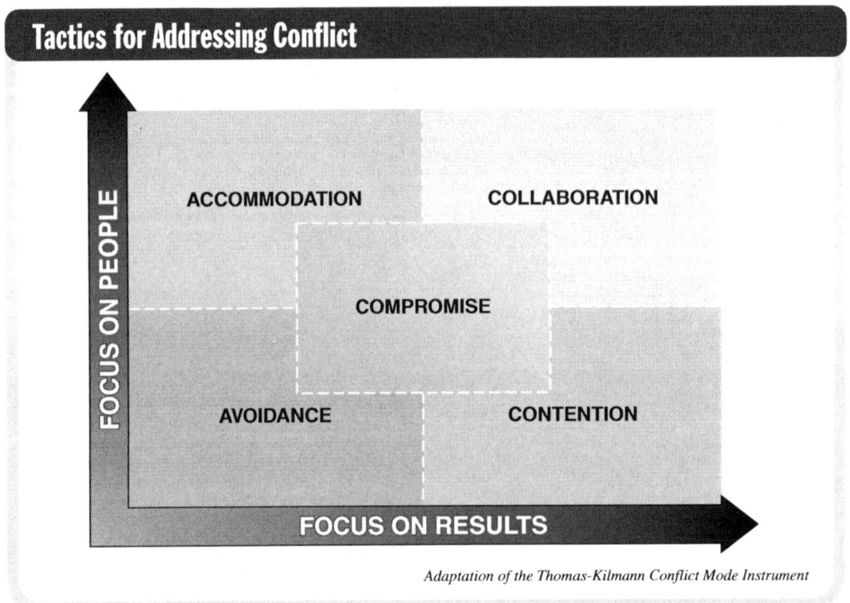

Figure 5-10. Whether the people involved in job-site conflicts are field leaders or crew members, they will likely view conflicts differently.

Damage To Another Trade's Installation

An effective field leader can use the advantages of each of these tactics to address conflict on the job site and also knows when it is best to either refer a conflict to upper management or to cut the company's losses. Addressing conflict in a private conversation as soon as it arises is the most effective way to keep a project on time and on budget. An effective field leader can use the following questions to help address conflicts:

- "What would you like to see happen, and can you describe it?"—This question helps the person describe the core problem. For example, a crew member may ask for respect. But until the person describes what respectful behavior is, the field leader will not know how to address the request—especially if the field leader has a different idea of how respect is shown.
- "What would it take to move forward, and how can we get there?"—This question helps the person describe specific steps. They may include an apology or a better understanding of the person's point of view before the person can move on and be a productive team member.
- "Are you willing to share the impact this has had on you, and are you willing to hear my perspective?"—This question moves the discussion from surface details to a working relationship. The point is so that both parties understand the effects of actions, assumptions, and language choices.

- "What ideas do you have that would meet both our needs?"—This question focuses the discussion on a resolution to meet the needs of both parties. It shows that the field leader is interested in a solution that is not selfish or egotistical.
- "What is of greatest concern in this situation?"—This question helps pinpoint the root issue, which is often based on a core value, like safety, belonging, or just making a difference. Often, the core value has been dismissed, disregarded, or trampled on by another worker or the field leader.

An effective field leader learns how to work with all types of people. A field leader is considered successful and earns respect when helping others reach their potential in business and in life. The work is both challenging and rewarding. Leadership skills take practice and time to master. A field leader who demonstrates effective leadership skills will have limitless opportunities and can become the type of field leader that others aspire to be.

Section 5.3 For Discussion

1. How can discrimination and harassment cause disruptive behaviors or conflicts on the job site?
2. Explain how bullies create a hostile work environment that is a breeding ground for conflict.
3. Explain why private face-to-face meetings should be used to address disruptive behaviors or conflicts caused by differing points of view, difficult workers, or difficulties with other trades.
4. What are five questions that a field leader can use to help resolve conflicts?
5. How can a field leader use different tactics to address conflict?

Section 5.3 Take Action

1. Role-play a field leader addressing disruptive behavior caused by differing points of view.
2. Role-play a field leader addressing a conflict caused by a difficult worker.
3. Role-play a field leader addressing a conflict with another trade.

Communication and Morale

Case Study

A large shopping mall is midway through completion, and the project has reached its peak of manpower. All the trades, from flooring to roofing and every trade in between, are currently working on site. Tensions are high as each trade maneuvers through the congested job site, trying to work as efficiently and productively as possible.

The electrical field leader is managing a crew of eight journeymen and two apprentices installing the lighting throughout the mall and parking lot. Some of his crew members are roughing in a new area. Others are beginning to trim out and test circuits in the nearly completed areas.

Thursday

8:00 am

The electrical field leader checks in with two crew members roughing in directional spotlights near the middle of the mall.

Crew Member 1:	"There's a discrepancy between the dimensions for the layout and the way it's been framed. We can't make the installation until the framing is reworked."
Electrical Field Leader:	"The framing field leader is a jerk, and I don't want to deal with him. Just get busy and move some of the lighting layout to fit the current framing."
Crew Member 2:	"Somebody on another crew mentioned that the directional lights are going to be used to highlight a large sculpture in the center of the mall. Won't moving the lighting layout affect that?"
Electrical Field Leader:	"Just get busy and leave the details to me!"

As the electrical field leader walks away, the crew members grumble about how they do not appreciate being blown off when they raised a legitimate concern. They also share their frustration with each other about being part of an incorrect installation just because their field leader does not want to coordinate with the framing field leader.

11:00 am

The electrical field leader walks outside to check on his crew members who are pulling wire for the parking lot lighting.

Crew Member:	"Did you read the response to this RFI before you handed it to me this morning?"
Electrical Field Leader:	"I glanced at it for a second. What's the problem?"

Case Study ...continued...

Thursday

11:00 am

Crew Member:	"There isn't an answer to the question about voltage drop and wire size. The electrical engineer responded that he couldn't read your writing, so he asked you to clear it up."
Electrical Field Leader:	"Great. We really needed that wire pulled this week. Now it's probably going to be another week until we get a response. Maybe I'll try and call the engineer and get an answer over the phone. In the meantime, the two of you find something to do and keep busy."

11:53 am

While walking back to the job trailer, the electrical field leader notices a couple of his crew members heading to the lunch trailer early. This is not the first time he has noticed them taking long lunches. So he decides to address the issue after lunch.

12:30 pm

Rather than talk to the two crew members who headed to lunch early, the electrical field leader decides to address the entire crew. He warns the crew that the next person he sees heading to the lunch trailer early is going to be written up and that this is everyone's verbal warning.

As the two crew members who are working on the spotlights head back to work, they complain about how frustrating it is to be lumped in with a couple of crew members who have been slacking off for months.

Crew Member 1:	"The field leader is a jerk! This would be pretty a good job if it weren't for the way he treats his crew. I wish I could get on one of the other crews."
Crew Member 2:	"Yeah, I don't get it. I get the same verbal warning about going to lunch early as those two fools? They've been doing that for months, and this is how he tries to resolve it?!"
Crew Member 1:	"I'm really starting to think he's in over his head. I hate working for guys like him!"

12:45 pm

The electrical field leader gets a call from a crew member who is trimming out fixtures at the end of the mall. The crew member states that there is a problem, and asks the field leader to come to the work area. When the electrical field leader arrives at the location, he finds that his crew cannot work because the flooring contractor moved the scissor lift they were working on and blocked them out by stocking a bunch of pallets in the area. The electrical field leader finds the flooring field leader to talk with him about moving the material so his crew can get back to work.

Case Study ...continued...

Thursday 12:45 pm

Electrical Field Leader:	"What do you think you're doing?! My guys were working on the lift you moved! They need to get back to work! You aren't even on the schedule to stock the job for two more days!"
Flooring Field Leader:	"We didn't have much choice because the delivery showed up early. I talked to the general contractor, and he said to bring it inside and put it in a place where no one was working. When I moved the pallets in, nobody was here."
Electrical Field Leader:	"Yeah, because it was lunchtime! The material has to go! You're holding us up!"
Flooring Field Leader:	"I followed the general contractor's instructions, and I'm not moving it again. I have to get back to work, so you're just going to have to deal with it!"

To get back at the flooring field leader, the electrical field leader instructs two crew members to find a couple of pallet jacks and scatter the pallets of material around the job site and to make them hard to find.

2:00 pm

The electrical field leader gets a call from the general contractor to come to his trailer immediately. When he arrives, he sees the two electrical crew members who were told to move the pallets. The flooring field leader and the general contractor are also there.

Flooring Field Leader:	"I came across these two knuckleheads putting one of my pallets up on the roof with a boom lift."
General Contractor:	"Do you want to explain what's going on here?"
Electrical Field Leader:	"First of all, I didn't tell them to put anything on the roof. I told them to move the pallets out of the way so they could work. But to be fair, those pallets shouldn't have been in the way in the first place."
General Contractor:	"This is completely out of line. If someone puts material in your way, I expect you to either work it out or come to me. I've informed your two crew members that I'm writing them up for this. I don't want any more of this kind of disruptive behavior."

The two crew members talk about their field leader as they walk back to the job site.

Case Study ...continued...

Thursday 2:00 pm

Crew Member 1:	"I told you we couldn't trust him! I can't believe he just stood there trying to cover his own butt while letting us take the rap."
Crew Member 2:	"We should have known this was going to happen if we got caught. Everybody I know warned me to get on anyone's crew but his."
Crew Member 1:	"I didn't want to believe that he was as bad as people said, but he just proved it. If he keeps it up, nobody in town is going to want to work with him.

3:30 pm

"One of the crew members who is working on the parking lot lighting approaches the electrical field leader to find out if he was able to get any information from the electrical engineer regarding the wire size.

Crew Member:	"Did you get the correct wire size for us?"
Electrical Field Leader:	"No, I got sidetracked and forgot to call. What have you guys been working on all day?"
Crew Member:	"Not too much because we thought you were going to be getting back to us soon. We just straightened up our tools and waited to hear from you. There really isn't much for us to do out there until we get that wire pulled."
Electrical Field Leader:	"You guys just stood around for four hours!? Are you kidding me!?"
Crew Member:	"It's not like you gave us any direction! And the last time I took the initiative to do something to fill in some time, you jumped down my throat for not talking to you first! We're getting mixed messages! What do you expect from us?!"
Electrical Field Leader:	"If you didn't have something to do, you should have told me! I can't read your mind! This whole crew is one of the worst I've ever had to deal with!"

As the job progresses, the electrical field leader continues to lose the respect of his crew as he ignores their concerns, blames them for his mistakes, and takes credit for the parts of the project that go well.

Two Months Later

The electrical field leader gets a call to meet with the general contractor and the architect in the mall to look at the lighting now that the sculpture is in place. When he arrives, he sees that they have a set of prints opened to the lighting layout for the directional spotlights.

Case Study ...continued...

Two Months Later

General Contractor:	"Take a look at the sculpture. We noticed some really strange shadows thrown across it by those spotlights. So we brought the prints to see if they were installed in the right place. They weren't."
Electrical Field Leader:	"We had to make some minor adjustments because the framing field leader didn't frame it correctly."
Architect:	"During the design of this project, we worked countless hours with both the sculptor and a lighting designer to make sure that these lights were not only the right type and size, but that they were also placed in the correct locations to cast the proper amount of light."
Electrical Field Leader:	"Like I said, I was just working with what I was given. Besides, I don't think anyone is going to notice."
General Contractor:	"Whether that's true or not isn't the point. If you knew the framing was off, why didn't you talk to the framing field leader?"
Electrical Field Leader:	"He and I don't see eye-to-eye, so I didn't feel like talking to him."
General Contractor:	"Then you should have come to me. That's what I'm here for. Here's what's going to happen now. I'm going to bring the framing field leader over here and have his crew cut open the drywall around those lights. Then I'm going to have him change the framing to accommodate the lights in the correct locations. Then your crew is going to remount the lights."
Electrical Field Leader:	"That's going to cost a lot of time and money! Are you sure you want to do that? Who's going to pay to reinstall drywall, finish, and paint it all again?"
General Contractor:	"The customer is going to get exactly what he asked and is paying for. Your company and the framing company can work out who is going to pay for the rework. You should have communicated with me that there was a problem, and this could have been handled without any cost to your company."

TAKE A CLOSER LOOK

1. How did the electrical field leader's poor communication skills directly lead to a loss of efficiency, production, and profit?
2. In what ways did the electrical field leader's actions cause his crew to lose respect for him?
3. How does this case study highlight the importance of effective communication skills?
4. How does this case study highlight the importance of crew morale to project success?

Case Study ...continued...

How Can an Effective Field Leader Communicate?

If the crew of another trade makes an incorrect installation that affects the installation:

- Listen to input from the crew.
- Put aside any personal differences with the other field leader.
- Tell the other field leader that a mistake has been made and collaboratively seek a solution.

When an RFI needs to be written:

- Clarify that all pertinent information is included before turning in the RFI.
- Make sure the RFI is legible and the question being asked is clear.
- Upon receiving a response, confirm the question has been answered before telling the crew.

If another crew stocks material in the way of an installation and refuses to move it:

- Explain to the field leader of the other trade why the material must be moved.
- If a resolution cannot be reached, ask the general contractor to address the problem.

PERSONAL PERSPECTIVE

1. Describe how effective communication skills could have been used to resolve a job-site conflict.

2. Role-play a conflict between two different trades and how it could have been resolved by the use of effective communication skills.

Case Study ...continued...

How Can an Effective Field Leader Sustain Crew Morale?

If two crew members leave for lunch early every day:

- Talk to the two individuals apart from the rest of the crew.
- Document the verbal warning and follow up if their behavior continues.

If crew members are in the line of fire because of something they were asked to do:

- Never ask crew members to do something that could lead to disciplinary action.
- Take responsibility for what crew members are asked to do.

If crew members are not productive while waiting for answers:

- Provide contingency tasks for crew members to do until answers are available.
- Listen to crew member suggestions regarding productive tasks to keep the project moving.

PERSONAL PERSPECTIVE

1. Describe how poor crew morale impeded work on a job site and how support from the field leader could have been used to help resolve the situation.

2. Role-play a crew support issue between a crew member and his field leader and how it could be resolved by both parties communicating effectively.

Case Study ...continued

How Can an Effective Field Leader Address Conflict?

If another trade places their materials in the way of an installation:

- Put personal differences with the field leader of the other trade aside to avoid conflict.
- Use the general contractor as a mediator if a resolution cannot be reached.

If crew members are in the line of fire because of something they were asked to do:

- Stand up for crew members.
- Take responsibility.

PERSONAL PERSPECTIVE

1. Describe how disruptive behavior impeded work on a job site and how the field leader could have used collaboration and compromise to help resolve the situation.

2. Role-play a conflict that occurred on a job site and how it could have been resolved by the use of effective conflict-resolution skills.

Job-Site Terms

A – B

addendum: A written document or drawing issued before the execution of a contract that modifies the original drawings and/or specifications.

alteration: A change or addition other than an ordinary repair or replacement.

authorized person: A qualified person approved and assigned to perform a specific duty at a specific location. Also known as a designated person.

bar chart: A schedule of work that uses a bar to represent each part of the job, with one end of each bar representing the starting date and the other end the completion date. Also known as a Gantt chart.

base bid: An original cost amount stated in a construction bid prior to adjustments for alternate bids.

base bid specifications: Construction specifications that describe the basic job, excluding alternates.

bid: An offer from a contractor to a potential client to construct a structure or to provide equipment that meets the conditions specified in the plans and specifications for a given price.

bidder: A person or contractor who develops a cost estimate and offer for a construction project.

bill of materials: A list of construction supplies needed to build a project that specifies individual quantities, descriptions, and sizes.

building code: Regulations adopted by a federal, state, county, city, or local government for the construction of buildings in a safe and structurally sound manner to protect the health, safety, and general welfare of those within or near the buildings.

building inspector: A building department official who determines if construction is in accordance with approved plans and applicable building codes.

building permit: A release form required by a state or local government that allows the construction of a permanent structure or alterations/additions to an existing one.

C – D

change order: A variation from original plans and specifications as directed by the owner or architect in written form.

code: A set of regulations governing building construction practices such as those for building, electrical, plumbing, and mechanical installations.

competent person: A person who is capable of identifying existing and predictable hazards in a work area and who has the authority to take prompt corrective measures.

completion: A point at which a building is ready for its intended occupancy or use.

construction contract documents: All legal and approved information concerning a construction project, including requests for bids, specifications, general conditions, prints, and addenda.

construction loan: A monetary arrangement between a lending agency and a building contractor.

construction site: *See* job site.

contract documents: All written and graphic materials used for the design and construction of a project, including the agreement, general conditions, prints, specifications, and addenda.

critical path method (CPM): A scheduling method that graphically shows the relationship between the parts of a construction project and identifies the jobs that must be completed before other parts of the project can start, as well as jobs that are independent of other jobs.

date of substantial completion: The date on which a construction project is ready for occupancy or use by the owner. Also known as the time of completion.

designated person: *See* authorized person.

design-build: A construction arrangement in which a contractor provides design work and building construction services.

E – I

employee: A person who works for another person or organization for wages.

employer: A person or organization that hires individuals to perform specific jobs for wages.

engineer/architect of record: The engineer or architect whose seal and/or signature appears on the plan drawings.

extra: Additional work or service added to a construction project after a contract agreement has been reached.

field order: A written statement from an architect to a contractor for a minor work change that requires an adjustment to the contract documents.

final acceptance: The complete acceptance of a project by the owner.

final payment: The completion of a contract payment from an owner to a contractor.

Gantt chart: *See* bar chart.

general conditions: Written agreements, included with a set of prints, that detail and describe building components and procedures.

general contract: An agreement between an owner and a general contractor for the completion of an entire construction project.

general contractor: A person or company that agrees to fulfill an entire building agreement and the various types of work to be completed. Subcontractors may be hired to do some of the work.

inspection: An examination of work by an appropriate authority to ensure it is done in accordance with plans and/or codes.

issuing authority: A governmental agency, office, or official with the authority to issue permits and/or licenses.

J – P

job conditions: A portion of a contract that describes the rights and responsibilities of contractors on a construction project.

job site: The property on which a structure is erected. Also known as a construction site.

joint venture: A project on which two or more people or organizations work together as partners.

licensed professional: A person whose services are required for a specific action by a governmental statute.

litigation: An engagement in a lawsuit using the judicial process.

negligence: A failure to use the required care or proper procedures for a particular operation.

negotiated work: Construction work assigned by an owner to a particular contractor or subcontractor without open bidding.

notice to proceed: A written statement from an owner to a contractor authorizing the beginning of work on a construction project.

overhead: A business expense, such as heat or electricity, that is not charged to a specific project.

overtime: Work performed beyond regular work hours, which is typically paid at a higher rate.

permit: A written license issued by an authorized agency.

progress chart: A graphic representation of the schedule and/or completion of the various phases of a construction project. Also known as a progress schedule.

progress schedule: *See* progress chart.

project: A planned-and-designed undertaking, such as the construction of a building.

project documents: All written materials relating to a construction project, including work orders, permits, requisitions, agreements, and other applicable documents.

project engineer: The person responsible for the design and management of the engineering aspects of a project.

project evaluation and review technique (PERT): A construction scheduling method in which individual events are charted and various portions of the project are anticipated as the job progresses.

project manager: The person managing the administrative and technical responsibilities of a project.

punch list: A list compiled by the owner and/or architect on a walk-through of a completed construction project that indicates the items that must be completed before finalization of the project.

purchase order: A written document of the details of a purchase, often coded with a specific number.

Q – Z

qualified person: A person in professional standing or authorized by an employer who has demonstrated capability of performing assigned work.

quantity survey: An estimation of amounts of materials and labor required for the construction of a structure.

quotation: An established price for materials and/or work in the form of a proposal.

record drawings: Prints that are marked to indicate changes to a project and provide a historical record of those changes.

rework: Correcting a defective, failed, or nonconforming item during or after an inspection.

special conditions: A section within contract documents that describes work unique to a particular construction project.

starting date: The time at which a project is approved to begin construction.

stop work order: A statement from an owner or inspection agency that work on a project shall not continue until certain conditions are satisfied.

subcontractor: A person or party that performs part of the work on a construction project under an agreement with the general contractor.

superintendent: A contractor's field representative responsible for supervision of an entire project.

take-off: A written list of types and quantities of materials from a print.

time of completion: *See* date of substantial completion.

unit price: The cost per item of material or work.

value engineering: The study of prints and specifications with the intention of making slight modifications to improve production and costs without changing the final product.

variance: A written authorization by an appropriate authority to deviate from standard building codes or ordinances.

volume estimating: The calculation of construction costs based on multiplying the total volume of a structure by a fixed cost per unit of volume, such as a fixed amount per cubic foot.

walk-through: An on-site inspection of a construction project by the owner, architect, and/or a representative of the contractor.

Index

Page numbers in italic refer to figures.

N

O

P

Q

R

S

T

U

V

W